Feeding of Non-ruminant Livestock

Feeding of Non-ruminant Livestock

Collective edited work by the research staff of the Département de l'Élevage des Monogastriques, INRA, under the responsibility of Jean-Claude Blum.

Translated and edited by Julian Wiseman
Lecturer in Animal Production
University of Nottingham School of Agriculture

Butterworths
London Boston Durban Singapore Sydney Toronto Wellington

All rights reserved. No part of this publication may be reproduced
or transmitted in any form or by any means, including
photocopying and recording, without the written permission of
the copyright holder, application for which should be addressed to
the Publishers. Such written permission must also be obtained
before any part of this publication is stored in a retrieval system of
any nature.

This book is sold subject to the Standard Conditions of Sale of
Net Books and may not be re-sold in the UK below the net price
given by the Publishers in their current price list.

Original French language edition © Institut National de
la Recherche Agronomique, 1984

English translation © Butterworth & Co. (Publishers) Ltd, 1987

British Library Cataloguing in Publication Data
Feeding of non-ruminant livestock
1. Animal nutrition
I. Institut national de la recherche
agronomique   II. Blum, Jean-Claude
III. Wiseman, Julian
636.08'4   SF95

ISBN 0–407–00460–2

Library of Congress Cataloging-in-Publication Data

Alimentation des animaux monogastriques. English.
Feeding of non-ruminant livestock

Translation of: L'Alimentation des animaux monogastriques.
Bibliography: p.
Includes index.
1. Animal nutrition. 2. Feeds. 3. Swine–Feeding
and feeds. 4. Poultry–Feeding and feeds. 5. Rabbits–
Feeding and feeds. I. Blum, Jean-Claude. II. Wiseman,
J. (Julian) III. Institut national de la recherche
agronomique (France). Département d'élevage des
monogastriques. IV. Title. V. Title: Non-ruminant
livestock

SF95.A475 1987   636.08'5   86-19268
ISBN 0–407–00460–2

Photoset by TecSet Ltd, Wallington, Surrey.
Printed and bound in England by
Robert Hartnoll (1985) Ltd, Bodmin, Cornwall.

Preface

Although the principal objective of this book was to provide a sound document in French dealing with the feeding of non-ruminant livestock, it was felt that the wealth of data relating to nutritional requirements and raw material composition accompanied by a thorough and comprehensive text ought to appeal to a far wider readership. Furthermore, in addition to the conventional livestock considered (pigs, broiler chickens, turkeys, laying hens, rabbits), there are valuable details relating to 'minority' species which are important in France but less so outside that country. However, the poussin, for example, has already made an impact in the UK and the guinea-fowl may soon follow. Certainly, it could be agreed that diversification in terms of species reared may be a particularly useful development in the livestock sector.

The book should appeal both to students of animal production/nutrition where the sections dealing with energy, protein, minerals and vitamins may be particularly appropriate, as well as to those involved in the commercial animal feed sector where the comprehensive data and comments on raw material composition should prove invaluable.

In France, it is customary to express energy-yielding value in terms of cal/kg. Accepting that the SI unit is in fact MJ/kg, it was felt that this latter term is not yet widely employed. Accordingly, the former is retained throughout the text.

Julian Wiseman
Nottingham, 1986

Acknowledgements

I would like to acknowledge the considerable assistance of B. Leclercq of Station de Recherches Avicoles, Nouzilly, Tours, France, my wife Lesley who gave invaluable advice on many of the finer points of French grammar, and Carol Stanton who typed the handwritten manuscript.

Contents

Introduction   xi
J. C. Blum

Part I General information on the feeding of non-ruminants

1. Intake, requirements, recommendations   3
   B. Leclercq, Y. Henry and J. M. Perez
2. Energy value of feeds for non-ruminants   8
   B. Leclercq, Y. Henry and J. M. Perez
3. Protein nutrition of non-ruminants   14
   M. Larbier
4. Mineral nutrition of non-ruminants   19
   B. Sauveur and J. M. Perez
5. Vitamin nutrition of non-ruminants   26
   J. C. Blum

Part II Dietary recommendations

6. Nutrition of growing pigs   39
   Y. Henry, J. M. Perez and B. Seve
7. Nutrition of breeding pigs   54
   Y. Henry, J. M. Perez and B. Seve
8. Nutrition of rabbits   63
   F. Lebas
9. Nutrition of rapidly growing broilers   70
   B. Leclercq, J. C. Blum, B. Sauveur and P. Stevens
10. Nutrition of laying hens   78
    B. Leclercq, J. C. Blum, B. Sauveur and P. Stevens
11. Nutrition of turkeys   86
    B. Leclercq, J. C. Blum, B. Sauveur and P. Stevens
12. Nutrition of guinea-fowl   95
    B. Leclercq, J. C. Blum, B. Sauveur and P. Stevens
13. Nutrition of ducks   102
    B. Leclercq, J. C. Blum, B. Sauveur and P. Stevens
14. Nutrition of geese   110
    B. Leclercq, J. C. Blum, B. Sauveur and P. Stevens
15. Nutrition of Japanese quail   113
    B. Leclercq, J. C. Blum, B. Sauveur and P. Stevens

| 16 | Nutrition of pheasants and partridges
B. Leclercq, J. C. Blum, B. Sauveur and P. Stevens | 116 |

Part III Composition of raw materials

| 17 | Comments on raw materials
D. Bourdon, C. Fevrier, B. Leclercq, M. Lessire and J. M. Perez | 123 |
| 18 | Tables of raw material composition
*D. Bourdon, C. Fevrier, J. M. Perez, F. Lebas, B. Leclercq,
M. Lessire and B. Sauveur* | 134 |

Index of feed ingredients   209

Contributors

J. D. Blum
Chef du Département de l'Élevage des Monogastriques,
Centre INRA de Tours-Nouzilly,
37380 Monnaie, France

D. Bourdon, C. Fevrier, Y. Henry, J. M. Perez and B. Seve
Station de Recherches sur l'Élevage des Porcs,
Centre INRA de Rennes-Saint-Gilles,
35590 L'Hermitage, France

F. Lebas
Laboratoire de Recherches sur l'Élevage du Lapin,
Centre de Recherches de Toulouse,
BP 12,
31320 Castanet-Tolosan, France

M. Larbier, B. Leclercq, M. Lessire, B. Sauveur and P. Stevens
Station de Recherches Avicoles,
Centre INRA de Tours-Nouzilly,
37380 Monnaie, France

Introduction

Domestic non-ruminant animals — definition and characteristics

The term *non-ruminant* has appeared as the opposite of the term *ruminant*. Thus any definition of the term requires some form of comparison.

In ruminants food is modified to a large extent by the rumen microflora before it is degraded by the animal's own digestive system. In non-ruminants this latter system operates initially; microflora only have a limited role, confined predominantly to the large intestine and based on the utilization of residues, both exogenous and endogenous, of digestion and absorption.

As food passes rapidly through the crop of birds, this only has a slight influence (amylolytic flora) which is in no way comparable to that of the rumen. To summarize, the species which are of interest are linked less by morphological characteristics (i.e. a simple stomach) than by a digestive physiology which differs from that of ruminants because it ascribes a major role to the digestion of food by the animal's own digestive processes and only a secondary role to microflora. The use of similar techniques in the rearing and feeding of these animals constitutes another common feature among non-ruminants of interest to animal science: for instance the provision of balanced compound feeds obtained from a mixture of raw materials, intensive rearing in large numbers, and rigid control of feed intake at certain periods of the life-cycle.

This is intended neither as a comprehensive work on nutrition nor as a collection of useful hints but rather is directed to those interested in the nutrition of pigs, rabbits and poultry who are looking for technical data necessary for the formulation of balanced compound diets and also for those who wish to adopt a dietary programme to meet changing requirements as and when they occur during rearing.

A basic knowledge of the digestive physiology and metabolism of non-ruminants is assumed. As man is himself a non-ruminant this knowledge is to be found widely in most classic works of physiology and biochemistry. On the other hand, we considered it necessary to review briefly some general facts about the nutrition of non-ruminants in Part I, with particular emphasis on the characteristic requirements as influenced by rearing conditions: to investigate optimum animal performance and, to an even greater extent, maximum economic output.

In the last two parts of the book, which form a technical dossier and which therefore are the most important, we have collated a considerable amount of new information not found in English publications and sometimes as yet unpublished.

Whether for nutritional recommendations (Part II) or for composition of raw materials (Part III) we have endeavoured to be as precise as possible by relying upon, in the first instance, our own results, then upon those of our fellow researchers and, finally, upon a literature review. In order to allow a choice relevant to individual economic situations we have sometimes indicated the possibility of using different feed levels which will promote varying levels of animal performance. For the same reason, we have also indicated the variability in raw material composition.

Our main objective has not been one of innovation. We do not propose to change the criteria of feed formulation. We have however had to choose between those currently employed and reject those whose use would be premature but which may nevertheless be of interest. We feel it is necessary to justify these choices.

Comments on criteria used in feed formulation

Within the limits of the food offered, in order to maintain itself the non-ruminant animal must obtain all essential nutrients in an available form. Thus the formulation must be complete and must account for both the animal's requirements and the characteristics of the feed (including both composition together with digestive and metabolic fate). For this reason we considered it important to bring together in the same book feed recommendations and characteristics of raw materials used in feed formulation. Among these characteristics, we are able to distinguish between those nutrients where values are easily determined by simple analyses, and those more difficult to assess because they are dependent both upon the feed and its utilization by the animal.

This being the case, energy value is expressed differently for each species or animal type (poultry values are applied to all birds in the absence of precise data for each species). In pigs and rabbits digestible energy has actually been determined and constitutes the most relevant term for balancing a diet. However, digestible energy does have disadvantages in so far as it overestimates the energy value of protein, especially the part not retained for synthesis and not completely catabolized for excretion via the urine. Bearing in mind this incomplete breakdown, metabolizable energy is a better measurement of the energy value of high-protein raw materials and thus allows for a better estimate of their relative worth in low-cost diets. However, metabolizable energy is not yet completely satisfactory as it places all nutrients on an equal footing to cover different energy needs (quantitatively and qualitatively) according to the physiological state of the animal and the composition of liveweight gain. All animal scientists would like to use an energy value which covers all losses and production requirements — net energy — in order to express requirements and the energy value of raw materials. Unfortunately net energy has the disadvantage of considerable reliance on the animal and its nutritional state. In this way, experimental conditions influence results and consequently authors present conflicting measurements. In fact there is not one but several different net energies for each type of production: in the case of the pig, for example, even given the availability of reliable data (which is not the case at present) net energy for fattening could not be validly used for the whole growth period and even less for other periods of the animal's life. Finally, the problem is made more complex by interactions between nutrients and metabolites, which results in non-additivity between different energy levels. All things considered, at present it appears impossible to construct a coherent system of net energy from the scattered and contradictory data found in the literature. The methods of expression adopted in this work have on the contrary the advantage of reliability. For pigs, both digestible energy (measured) and metabolizable energy (calculated) have been mentioned — the latter may be preferable for determining relative worth.

The requirements and raw material composition of amino acids are presented without reference to their availability. It is however known that there are a number of factors influencing this — some linked to the animal (species, age) others to the feed (origin of raw materials, processing). Unfortunately measurements are difficult and associated with errors such that it is impossible at present to distinguish reliably between different raw materials.

In fact, nutritional recommendations take this variability into account by adequately overestimating the necessary quantities. The saying 'he who can do more, can do less' can only too well be applied to nutritional problems. We cover ourselves by oversupplementation (with the unlikelihood of creating an imbalance through excess — methionine in rabbits), but we end up with costly waste. This is the case for protein and essential amino acids. Any deficiency is detrimental but any excess is broken down without much gain to the animal (low energy value). All uncontrolled variable factors increase the necessary safety margins. In this respect the unavailability of amino acids is an important factor. Without doubt we will one day express nutritive values in terms of available amino acids. At the same time we will better estimate the quality of different protein sources and be able considerably to reduce protein levels in diets. Important applied research in this field is continuing for all non-ruminants.

What is valid for protein is equally so for minerals and vitamins. They are not completely available and factors influencing their availability are still poorly understood. However, it seems possible and useful in this context to take into consideration phosphorus for poultry: requirements and composition of raw materials will be expressed in terms of available phosphorus.

The omission of any comments and data on feed additives which are widely used for non-ruminants will certainly not pass unnoticed. We felt that inclusion of such information, which is governed by legislation, would be superfluous. Moreover, such additives are more often used for therapeutic than for nutritional reasons whether by chemical or auxiliary means. It is generally known that growth factors, anti-stress agents and others have a greater effect when rearing conditions are not favourable.

Conclusions

In this work our intention has been to inform rather than to confuse when considering the concepts employed in animal nutrition. We have refused to give way to elitism by adopting systems which have yet to be put to the test (net energy). We have however taken into consideration our most recent research work; each contributor was asked to place his own work within the context of a review of the literature.

In the desire for clarity, we have run the risk of being rapidly overtaken. This is therefore only a first edition which we hope will be revised subsequently. This will then not only provide an opportunity for updating, but will also allow for corrections and omissions. We hope therefore that many of our readers will offer constructive criticism. All remarks and suggestions will be welcomed to improve the content of subsequent editions. The promotion of wide discussion on rearing of animals is necessary given that in the area of animal nutrition, as elsewhere, participation is better than *laissez-faire*.

Acknowledgements

This is a collective work, the writing of which called upon the services of many researchers and technicians from the Département de l'Elevage des Monogastriques. The list of principal authors at the beginning of this book omits many other contributors. I want to stress that B. Leclercq played a key role in organizing meetings and following the progress of the different authors; furthermore he reviewed all poultry publications. The role of the development officers should be emphasized: D. Bourdon and J. M. Perez (Station de Recherches sur l'Elevage des Porcs — SRP), M. Lessire (Station de Recherches Avicoles — SRA). Y. Henry and F. Lebas took charge of the chapters on pigs and rabbits respectively. B. Sauveur had the thankless job of editing the text; he was helped in this long and difficult work by the administrative staff at the SRA: Y. Salichon and M. Plouzeau.

J. C. Blum
INRA, Monnaie, 1986

Part I
General information on the feeding of non-ruminants

Chapter 1
Intake, requirements, recommendations

Animals must obtain from their feed all nutrients necessary to permit the renewal of tissues, growth (during fattening and gestation) and the synthesis of the constituents of milk and eggs. The quantities of nutrients assimilated during all of these activities define requirements: requirements for water, dietary energy-yielding ingredients, protein and essential amino acids, minerals and vitamins. Requirements vary according to both the physiological condition of the animal and its health status. The non-productive adult is considered to be at maintenance. Additional requirements associated with development or production are frequently accounted for by a simple addition to maintenance. Total daily requirements for a young fattening animal may for example be broken down into maintenance requirements associated with a given liveweight and a production requirement based on daily liveweight gain.

The production of animals may be precisely discussed when requirements are known, and those factors which modify them understood, by starting with feed intake and appreciating those elements which influence digestibility and metabolism. It then becomes possible to define the characteristics of the feed to be given.

However a certain imprecision always exists in the determination of requirements, in the knowledge of the exact nutritive value of raw materials and the estimate of the quantity of feed consumed. In addition, the production of compound feeds and their handling may be accompanied by a certain heterogeneity of the mix.

For all these reasons a margin of security is adopted during formulation designed to ensure that animal requirements are met. It is thus appropriate to move from a consideration of requirements to one of recommendations. In that which follows, the principal rules governing the nutrition of non-ruminants in the context of intake, requirements and recommendations will be considered.

Regulation of feed intake

Although there may be complex interactions between requirements and intake, it is possible to break up those factors which influence the latter into those associated with the feed and those associated with the environment (primarily temperature).

Influence of the feed

Non-ruminants are able to control their intake almost completely so that their energy requirements are covered. An increase in dietary energy concentration is accompanied by a corresponding reduction in feed consumption such that metabolizable energy intake varies little. This degree of regulation is species specific and will be considered for each one in the relevant chapter.

Intake is influenced to a lesser extent by dietary protein levels. In the case of low levels some species tend to increase feed consumption such that amino acid intakes are maintained. With excess dietary protein, on the other hand, a slight reduction in feed intake is observed but without a change in growth rate. This situation is particularly apparent in the chicken and allows to a certain extent a degree of control over growth.

There are other specific dietary factors which may have a small but significant effect upon intake. Thus the laying hen will respond to dietary calcium during egg-shell formation. If calcium is evenly and finely distributed throughout the diet, the hen will overeat in order to consume sufficient calcium. On the other hand, if calcium carbonate is present but distinct from the remainder of the diet then this will allow a corresponding reduction in consumption of the latter.

The form of presentation of the feed may also play a role in certain species: in particular, pelleting increases feed intake (especially if dietary energy levels are low) in chickens and ducks but has little effect with other species. Details are given concerning this under each production system.

Finally, feed intake may be reduced by the presence of factors that influence palatability or which may be toxic. The former are comparatively rare in birds but are found more frequently in pigs and rabbits.

Influence of ambient temperature

An increase in ambient temperature is accompanied by a reduction in feed intake, in a more or less linear way, from low temperatures up to the thermoneutral zone. Frequently the reduction in dietary energy intake is at a rate slightly faster than the lowering of energy requirements; the animal is therefore in a progressively lower state of positive energy balance with increasing ambient temperature and this is often associated with a slower growth rate. Above the thermoneutral zone appetite falls rapidly and the animal finds itself in a greater state of nutrient deficiency. This deficit is one of the causes of reduced performance in hot climates. In order to remedy this situation, albeit only partially, the dietary energy level may be increased. Parameters which allow the calculation of changes in energy consumption as a function of ambient temperature will be given for the commoner species in subsequent chapters.

Feed allowances

Daily feed intake may be restricted to a predetermined level lower than that which would be consumed voluntarily. In practice this is found in fattening pigs, pregnant sows, the pullet and the majority of poultry breeders.

Control of the level of intake allows control of the rate of fattening and an improvement in both fertility and, frequently, the health of the animal; it allows, additionally, the fine adjustment of dietary nutrient levels other than energy such that requirements may be more precisely met. Finally it is an important means of economic management. Feed intake may be calculated as a function of animal size (as well as age), genotype and ambient temperature. Precise guides to permit the fixing of daily intake corresponding to each species or to performance criteria supplied by breeding companies will be given in subsequent chapters. It is hoped that, following the use of such advice designed for a specific level of performance, a precise quantity ingested in this way will not result in varied output.

Concept of requirements
Requirements of an animal

The principal requirement of an animal is associated with its energy expenditure. Thus, after water, dietary energy-yielding constituents are those that most rapidly influence the health and survival of the animal if withdrawn. Additionally, energy requirements are the most sensitive to surrounding conditions and, as has already been discussed, the most important determinants of feed intake. The differences in feed intake between animals are explained largely by their respective energy requirements.

The distinction between energy requirements for maintenance and production is largely theoretical rather than physiological. Thus it is extremely rare to find feeding situations corresponding to the concept of maintenance where the animal is in a state of equilibrium without gain or losses of fat or protein. Such a situation may be found in adults but almost never in the young animal. However statistical models, both linear and non-linear, estimate with a good degree of reproducibility energy expenditure corresponding on the one hand to synthesis (fat and protein associated with production) and, on the other hand, to maintenance. The latter is based on liveweight or metabolic liveweight (liveweight$^{0.75}$).

Energy requirements based on production depend to a large extent upon the composition of such production. Thus, the higher it is in fat then the greater are energy costs as adipose tissue contains only small amounts of water. On the other hand protein synthesis, which characterizes living tissue, is associated with large amounts of water (at least 75% of muscle) and the energy costs of such production are accordingly lower.

Energy requirements of production are on the whole independent of surrounding conditions but on the other hand are strongly influenced by genotype. Maintenance requirements are affected considerably by environment: in recommending values for each species at a given rearing temperature coefficients will also be given to allow the calculation of values with changes in environmental temperature. In addition, calculations to predict energy requirements from both ambient temperature and characteristics of the animal are presented in the next chapter (pages 12 and 13).

Other requirements of the animal (protein, minerals, vitamins) may also be separated into those for maintenance and those for production. In the majority of cases (particularly for protein and amino acids) the latter are considerably more important. A requirement expressed in terms of the quantity of a nutrient per animal per day is dependent to an extent upon animal size but, more importantly, on its level of production and, therefore, upon genotype. On the other hand, this requirement is largely independent of environmental conditions (temperature). Obviously this would be the opposite if the requirement were to be expressed in relative terms: as a concentration in the feed.

The different effect of environment on the requirements for energy and requirements for other nutrients means that the characteristics of feeds need to be modified as a function of rearing conditions. With high ambient temperatures (during summer or under tropical conditions) requirements for a nutrient expressed as a percentage of the feed will be more important than under more temperate or colder conditions: this explains to a degree the variability in tables of requirements from different countries or organizations.

Generally, conditions which promote a reduction in feed intake must be accompanied by an increase in the level of protein, minerals and vitamins in the feed. Recommendations for each species are presented in terms of absolute values (quantities for a given length of time) and in terms of percentages of the feed for a level of performance at a given level of intake and ambient temperature. This will allow farmers or feed compounders to modify their feed according to the conditions operating (temperature, genetic potential of the animal).

Requirements of a population

The nutrition of non-ruminants is studied more and more in terms of the group rather than the individual animal. It is therefore necessary to consider the variability in performance and intake. This has not yet been the subject of much research but must, however, be responsible for some of the differences between tables of requirements or recommendations.

All factors which introduce variability in performance will increase requirements. The requirements of the best animals must be met even if this results in wastage associated with those whose performance is mediocre. Any attempt designed to reduce variability in genetic potential or environmental conditions will, for the same mean level of production, result in a lowering of requirements. Ultimately the decision is an economic one: it is a question of knowing that proportion of a population whose requirements are met beyond which level no benefits would accrue were there to be any additional expenditure.

Practical recommendations

There are many reasons why margins of security are important for farmers and feed compounders when considering requirements, and these are considered in the following five principal categories.

(1) *Uncertainty associated with raw materials* Frequently, the chemical composition of raw materials under consideration is not precisely known, either as a consequence of natural variability or imprecision concerning measurement. Moreover the biological availability of nutrients may be unknown or variable.
(2) *Variability in the compound feed* Precision in weighing or the homogeneity of mixtures are frequently limited by the equipment used. Moreover conditions of handling and storage are associated with a degree of variability in feed composition which is often difficult to assess.

(3) *Uncertainty associated with the exact nature of requirements* Knowledge of requirements is frequently imperfect for the reasons outlined previously. Thus the species under consideration may have been the subject of too few studies, or may show a wide variability in performance. Finally, there frequently exists considerable uncertainty regarding the requirements for nutrients hitherto considered as secondary: this is the case for example for all essential amino acids other than lysine and those containing sulphur.

(4) *Uncertainty related to knowledge of feed intake* Considerable risks of under-consumption exist as a consequence of environment (seasonal variation in temperature, stress), the diet (higher dietary energy levels than anticipated, factors influencing palatability) or indeed genotype.

(5) *Influence of disease* A reduction in hygiene levels during rearing may lead to an increase in requirements under consideration, under-consumption, reduction in growth rate and finally a modification in the digestion and metabolism of feed.

It is by taking these risks into account and adopting a margin of security that satisfactory performance is guaranteed. The recommendations provided only make allowances for uncertainty associated with inadequate knowledge relating to requirements, with the precise level of intake and with the availability of nutrients. Those using this book will be able to adopt higher values based upon their own special conditions of rearing and feed technology. It must be stated however that any effort to reduce sources of error will lead to important savings.

Chapter 2
Energy value of feeds for non-ruminants

Definition of terms used to express energy

The general scheme describing the utilization of energy-yielding nutrients in non-ruminants is presented in *Figure 1*. Apparent digestible energy (ADE) is obtained from the difference between gross energy (GE) of the feed consumed and the GE of the faeces. Apparent metabolizable energy (AME) is obtained from the difference between GE of the feed consumed and the total energy excreted in faeces, urine and gas (methane and hydrogen). Metabolizable energy thus defined may also be referred to as 'classical' or 'apparent uncorrected'; it corresponds to the energy available for the metabolic needs of the animal, being maintenance and production. It forms the basis of energy measurements and calculations of net energy.

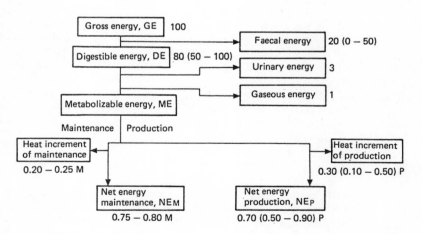

Figure 1 Partition of dietary energy and relative value of measurements in non-ruminants.

Net energy (NE) corresponds to the difference between metabolizable energy and heat losses associated with the utilization of metabolizable energy. It may be subdivided according to how it is utilized: NE for maintenance encompasses a basal metabolism (animal at rest, not eating and within its thermoneutral zone) as well as the costs of thermoregulation and those associated with physical activity; NE for production corresponds to the energy contained in body tissues and the various products (milk, muscle, fat, eggs).

Measurement and applicability of each term

Gross energy (GE)

This may be directly measured using bomb calorimetry. It may also be determined by regression if the chemical composition (by the Weende analysis) of the feed is known.

$$GE = 57.2\ CP + 95.0\ EE + 47.9\ CF + 41.7\ NFE + \Delta i$$

where:

    GE = gross energy (kcal/kg)
    CP = crude protein (%)
    EE = ether extract (%)
    CF = crude fibre (%)
    NFE = nitrogen-free extract (%)
    Δi = correction factor for certain raw materials

Digestible energy (DE)

In poultry, anatomical features (the cloaca) prevent the easy separation of faeces and urine and make it difficult to measure digestible energy. On the other hand, this measurement is straightforward and precise in pigs and rabbits.

Digestible energy is only marginally influenced by level of intake. Its major drawback is that it overestimates the energy value of protein since it does not take into account the incomplete oxidation of amino acids by the animal (associated with the excretion of urea and uric acid). The DE values for pigs of raw materials with a mean defined chemical composition that are given in the tables are compilations of experimental results. For certain raw materials (cereals, fibrous by-products) these values may be corrected by the use of equations that take into account the level of crude fibre determined in a sample. Such a correction is not applicable to a mixture of raw materials.

Metabolizable energy (ME)

In poultry, ME is the easiest energy value to measure and, consequently, is at the moment the most commonly used term. Classical ME (or apparent uncorrected) may be corrected on the basis of nitrogen retention which allows for comparisons between animals at different physiological stages. The correction is, depending upon author, between 8.22 and 8.73 kcal/g nitrogen retained in birds: the adult at maintenance (cock) does not retain nitrogen (no correction factor) which therefore allows simplified measurements (Chapter 17).

In addition ME may also be corrected by deducting unavoidable losses not directly related to the feed ingested (endogenous faecal and urinary losses). This gives true ME which is some 5–10% higher than classical ME. This difference between the two expressions of ME is greater the lower the level of feed consumption (unpalatable feed).

True ME is a means of expression which may occasionally improve precision and allow a better comparison between raw materials. Its determination is however less easy than with classical ME. It is possible, as with classical ME, to correct true ME on the basis of nitrogen retention.

Measurements of ME have the advantage of not having to rely upon complex equipment, as long as gaseous losses (from intestinal fermentation) are ignored; these latter may however be important in pigs. The principles of additivity generally hold in the case of ME. The ME value of a mixture is essentially the sum of the values of the constituents. ME provides a better estimate of the energy value of protein than DE.

In the absence of direct determinations, ME may be estimated using regression equations based on chemical analyses. These equations are generally not very precise because they are based on chemical analysis without taking digestibility into account. Some equations have been proposed that may be applied to all raw materials and others specifically for groups of raw materials (barley, meat meals). The latter are unfortunately of limited value and sometimes lead to results with cereals, for example, which are not easily repeatable in time (effect of year, technological advances, breeding of new genotypes).

For poultry, the following equations may be used for compound feeds:

(1) ME (kcal/kg) = 35.2 CP + 78.5 EE + 41.0 S + 35.5 Su
(2) ME (kcal/kg) = 36.1 CP + 76.9 EE + 40.6 S + 26.1 Su
(3) True ME (kcal/kg dry matter) = 3951 + 54.4 EE − 88.7 CF − 40.8 ASH

where:

CP = crude protein (%)
EE = ether extract (%)
S = starch (%)
Su = sugar (%)
CF = crude fibre (%)
ASH = ash (%)

For pigs, apparent ME is used assuming that the diets are balanced in terms of protein and amino acids. In order to avoid an overestimation of the energy value of protein-rich raw materials ME values obtained from directly determined DE values may be used. To achieve this, an average figure for urinary losses, usually obtained when the dietary levels of amino acids are balanced (resulting in a value of 50% nitrogen retained as a proportion of nitrogen digested) and methane energy losses (representing 1% of DE) are taken into account. The following equation is thus obtained:

(4) ME/DE × 100 = 99 − 0.07 (DP/DE)

where:

DP = digestible protein (g/kg)

The ME values for pigs in the tables of composition given were derived in this way.

Net energy (NE)

The NE value of a feed is obtained from the ME value by multiplying it by the overall coefficient of utilization (k) which includes the utilization for maintenance (k_m) and for production (k_p). The value of the latter is dependent upon the chemical composition of production — the relative concentration of fat, protein and, for milk, carbohydrates, which are deposited. Taking this into account, the coefficient k is also dependent upon the relative importance of maintenance and production as well as the composition of production. It is easy to appreciate therefore that the NE value of a feed is dependent to a large extent upon the animal which consumes it (its age, body composition, etc.) in contrast to ME and DE: following from this, NE is a better indicator than the latter of the utilization of dietary energy by the animal.

The value of k_m varies between 0.7 and 0.8. It is slightly higher for carbohydrates than for fat (+5%) and lower for protein (−20%). The efficiency of utilization of ME for protein deposition (kcal of protein/kcal ME required) is in the region of 0.5 for tissue protein and 0.65 for egg protein. That for fat deposition is dependent upon the carbon source; the following values may be adopted: 0.60 for dietary protein, 0.75 for carbohydrates, 0.85 for fat and 0.60 for volatile fatty acids.

The current methods available for calculation of the NE of feeds for the pig from the proportion of digestible nutrients in compounds or raw materials are:

(1) *The NEF system* (Rostock, East Germany) This is based upon numerous experimental results. It only takes into account the efficiency of fat deposition and applies this to total requirements (maintenance and production). The general prediction equation is:

NEF = 2.49 DCP + 8.63 DEE + 1.5 DCF + 3.03 DNFE

where:

NEF = net energy (kcal/kg)
DCP = digestible crude protein (g/kg)
DEE = digestible ether extract (g/kg)
DCF = digestible crude fibre (g/kg)
DNFE = digestible nitrogen-free extract (g/kg)

(2) *The system of Just* This is based upon the correction of ME by a constant term and thus takes into account variations in the utilization of energy for growth associated with the dietary energy concentration of the feed:

NE = 0.75 ME − 450

where both ME and NE are expressed as kcal/kg dry matter.
 The energy value is expressed in units of barley equivalents where 1 unit of NE is 1850 kcal/kg barley in the unprocessed state.

Criticisms of the systems and proposals

The concept of NE, admirable though it may be in better encompassing the metabolic processes that involve dietary energy, presents several disadvantages however. Its direct measurement requires expensive equipment. The distinction between maintenance and production raises problems of estimation and generalization. All the modifications proposed above therefore carry with them a degree of simplification of the problems: estimation of the losses associated with heat production, and the role of the efficiency of lipogenesis within overall energy requirements. In addition, a system based on NE does not fulfil the conditions of additivity which is the basis of feed formulation: such a system will frequently tend to give an advantage to dietary fat and penalize protein-rich sources. Consequently compound feeds formulated using the system will tend to be associated with excessive fattening in animals. Net energy is difficult to measure and has an inadequate basis, both experimental and theoretical; it cannot therefore be recommended for the practical estimation of feed destined for non-ruminants.

On the other hand, DE is much easier to measure in pigs and rabbits. It fulfils the conditions of additivity but leads to an overestimation of the value of protein.

For feed destined for birds, ME is undoubtedly the most appropriate system because it allows the most precise and rapid measurements. The principle of additivity is met and this allows the use of linear programming. It is proposed to use classical ME corrected to zero nitrogen retention in adults, to a nitrogen retention of 30% for laying hens and finishing birds, and 40% for growing birds: the correction factor used is 8.22 kcal/g retained.

For the pig the use of apparent ME would not result in the overestimation of protein sources. However, for the moment DE is preferred because of the low number of results available for ME.

Calculation of energy requirements

Many equations have been proposed to allow the calculation of energy requirements of animals. Only those relevant to birds will be given as these are the ones where most progress has been made.

Growing birds

The relevant equation is:

$$ME = [105 + 4.6 (25 - T)] W^{0.75} + 10.4 F + 14.0 P$$

where:

- ME = energy requirements (kcal ME/day)
- W = liveweight (kg)
- F = fat deposition (g/day)
- P = protein deposition (g/day)
- T = ambient temperature lower than 25°C

Below 25°C, energy requirements follow different patterns according to species, feathering and the quantity of subcutaneous fat. The equation applies above all to the chicken but nevertheless allows a correct estimation of the requirements of other species of poultry under normal rearing conditions (close to the thermoneutral zone).

Laying hens

The best equations are those of Emmans:

$$ME = (170 - 2.2T) W + 5 \Delta W + 2 E \quad \text{(Leghorn)}$$
$$ME = (140 - 2T) W + 5 \Delta W + 2 E \quad \text{(Rhode Island)}$$

where:

ME = energy requirements (kcal ME/day)
ΔW = mean liveweight gain (g/day)
W = liveweight (kg)
E = egg weight produced (g/day)
T = ambient temperature (°C)

These equations allow initially an estimation of the energy requirements of hens but do not take into account large differences which may exist between individuals based on feathering or physical activity (battery or free range).

Chapter 3
Protein nutrition of non-ruminants

Concept of amino acid essentiality

Protein synthesis in animals requires the simultaneous presence of around twenty amino acids. Some of these may not be synthesized by the organism or are synthesized at a rate insufficient to meet its requirements: these are named essential. A second category includes those which are strictly non-essential. Finally there are those that are semi-essential which may be synthesized from essential amino acids: this is the case with cystine and tyrosine which may be synthesized from, respectively, methionine and phenylalanine.

This classification is based upon metabolic processes: it is dependent upon species as well as physiological state. Before considering amino acid metabolism in slightly more detail it is worth mentioning that amino acids provided in excess may not be stored for any length of time: they are catabolized or excreted. On the other hand a non-essential amino acid may be the limiting factor in growth if its level in the diet is insufficient and if the essential amino acids responsible for its synthesis are similarly only present in marginal amounts. There are therefore a number of facets associated with the concept of amino acid essentiality and all must be taken into account during feed formulation.

It is now proposed to consider briefly the metabolic pathways that involve amino acids and to outline those principal factors which influence the nutritional value of protein feedstuffs. Protein requirements for each type of production system will be considered in subsequent chapters.

Amino acid metabolism in non-ruminants

Biosynthesis

Amino acid biosynthesis is only possible if the organism has available amine groups ($-NH_2$), carbon skeletons and enzymes specifically responsible for transamination.

Whereas the amine groups arise from the degradation of other amino acids, carbon skeletons may be present as intermediaries of carbohydrate metabolism (α-keto derivatives).

Only two amino acids (lysine and threonine) may not be synthesized from their respective α-keto precursors because the corresponding transaminases do not exist in higher animals. This explains their essentiality.

Amino acid degradation

In a state of starvation, or when the provision of dietary energy is insufficient and accompanied by low glycogen reserves in the body such that glucostasis may not be maintained, certain amino acids are degraded and their carbon skeletons converted into glucose (gluconeogenesis). The metabolic reactions involved may additionally lead to the production of compounds not involved in energy metabolism such as hormones and other chemical mediators: thyroxine, and adrenaline and dopamine, for instance, arise from phenylalanine and tyrosine, respectively.

When dietary levels of amino acids are higher than requirements for protein synthesis, the excess is catabolized. In pigs and rabbits, urea represents the principal means of nitrogen excretion. Uric acid plays the same role in birds because the urea cycle does not exist. In birds uric acid synthesis is controlled by hepatic xanthine oxidase whose activity is increased with higher levels of dietary protein. The pathway commences with one molecule of glycine which explains the relatively higher requirement for this amino acid in birds; their synthesis of glycine may be insufficient to meet both the requirements for growth and uric acid synthesis. In this respect serine may be used in the synthesis of glycine and may replace it in the diet.

Under normal physiological and nutritional conditions blood uric acid levels vary between 30 and 100 mg/l, and the quantity of uric acid excreted daily is 4–5 g. An abnormal increase in the blood level of uric acid will result in its precipitation in the kidney, joints, pericardium and elsewhere. Such a situation is promoted by the consumption of feed with excessive levels of protein or a deficiency in vitamin A.

Nutritional value of protein feedstuffs

Concept of amino acid availability

The factors to be considered when discussing the efficiency of protein utilization can be classed into two groups. Firstly, external factors are related to rearing conditions: method of feeding, level of intake, levels in the feed (of energy, vitamins, minerals), temperature, etc. The study of these factors allows the definition and expression of nitrogen requirements taking into account in addition daily feed intake and dietary energy levels.

Secondly, internal factors are concerned with protein itself. The nutritional value of protein may be estimated with reference to the percentage of nitrogen ingested that is retained for protein synthesis. This is dependent, *a priori*, upon the amino acid composition of raw materials but the relationship may not be direct if the protein has been subjected to processing or a long period of storage. In this case the level of amino acids determined by simple measurements does not correspond to the level of available amino acids.

The availability of an amino acid may be defined as the percentage utilized for protein synthesis when the amino acid in question is the only limiting nutrient in the diet. There are two consequences following from this definition: the first concerns the methodology used and the second is related to those amino acids for which there are problems of availability.

Given the direct relationship with the level of protein synthesis, availability may be equally well determined in the young (growth anabolism) as in the adult (production of milk, eggs) or at maintenance (renewal of tissue protein). The application of these methods will be considered subsequently.

Availability is only important for those amino acids which may be limiting in the diet. Thus lysine is of particular importance because it is strictly essential, it is generally at low levels in the majority of feed proteins (cereals, meals with the exception of soya-bean) and because it has an ϵ-amine group which is prone to reaction with carbohydrates and lipids.

Finally it should be remembered that during heat treatment and throughout the storage of protein feeds, many reactions may take place within raw materials. Having taken part in these reactions (of which Maillard reactions are of particular relevance) amino acids are not normally released by enzymic hydrolysis in the digestive tract and are therefore rendered unavailable for protein synthesis. The resultant reduction in availability under certain conditions may be of the order of 50%.

Measurement of amino acid availability

The methods used may be classified into two groups. The first consists of chemical techniques which do not rely upon the use of animals (method of Carpenter) and provide a direct measurement of availability in raw materials by using chemical reactions of varying specificity. The second group includes all those methods where availability is measured with reference to *in vitro* or *in vivo* transformation or utilization arising from the protein in question provided at a given level (*Figure 2*).

Generally, chemical methods are suited for the determination of available protein in animal protein feedstuffs. They are not appropriate for vegetable protein raw materials high in carbohydrates.

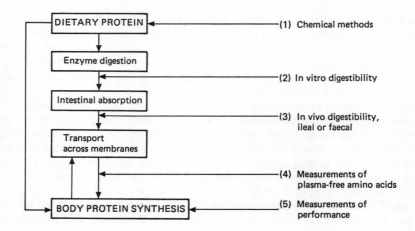

Figure 2 Assessments (1–5) of availability of limiting amino acids. Methods are associated with various areas, from the food to tissue protein.

In methods based on growth assay, the ability of a protein to replace a specific amino acid during the growth of a young animal is measured. The methods differ in terms of the measurement selected: liveweight gain, protein gain, food conversion ratio. In theory, they obviously lack specificity since the performance measured is not solely due to the dietary level of the available amino acid under consideration but is also related to all the factors which influence the efficiency of protein utilization. In practice, such methods are costly because at the same time they require a large number of animals and a long experimental period. However they remain the only definitive practice and direct means of verifying the values for availability obtained by all the other methods.

In methods based upon provision of free amino acids, availability may be clearly defined either by the amount of amino acids absorbed and extracted from portal blood or by the amount present in peripheral blood and tissues when requirements for protein synthesis have been met.

If it is accepted that the peripheral blood or tissue concentration of free amino acids is based upon the quality and quantity of ingested protein, it is dependent additionally on the rate of anabolism. Furthermore many factors may influence the concentration of free amino acids: method of feeding, time of sampling of blood or tissue, nature of dietary energy components, genotype and individual animal variability. The multitude of factors means that, for each type of raw material, it would be essential to adopt a methodology which employs the same diets identical in all respects except for the amino acid whose availability is under study. This problem, as well as the need to have a number of experimental levels to improve accuracy, precludes the adoption of these methods for systematic use.

The methods based on digestibility are concerned with the measurement of the percentage of amino acids liberated by enzymic hydrolysis (*in vitro* methods) or liberated by digestion and absorbed by the intestinal mucosa (*in vivo* methods).

In vitro digestibility, although representing an advance over methods that involve the simple measurement of amino acids in the estimation of the nutritional value of a protein, only imperfectly reproduces the conditions involved in the physiology of digestion. It may serve as an indicator of enzymic digestion but only gives little information relating to intestinal absorption or to the utilization of amino acids for protein synthesis.

Methods involving *in vivo* digestibility are a better approach to the problem. Availability is measured by reference to the percentage of amino acids absorbed either throughout the digestive tract or as far as the ileum in order to avoid interactions with microflora in the large intestine. The methods therefore require the measurement of the amino acid in the raw material under consideration, and again in the faeces or ileal contents.

In birds the problem is complicated by the mixing of urine and faeces via the cloaca: their separation requires surgical techniques, or the use of chemical methods which allow the differentiation between faecal and urinary nitrogen.

In all species of mammals and birds the use of digestibility as a means of estimating availability does not take into account the kinetics of digestion or of passage of digesta. It neglects in addition the role of intestinal microflora (in the case of the measurement of overall digestibility) and, more generally, all factors modifying the physiology of digestion which influence amino acid availability.

The measurement of digestibility (total or ileal) ought nevertheless to constitute a useful criterion for availability in the future and it is possible to foresee tables of analysis of raw materials containing mean values for the digestibility of the different amino acids. The data obtained up to now, however, are only fragmentary and associated with the same uncertainties of methodology as with the determination of dietary energy values (composition of experimental diets, method of feeding, conditions of faecal collection).

Practical consequences

Estimation of requirements

Knowledge of the metabolic peculiarities of each non-ruminant species allows decisions on the nature of the essentiality of amino acids and the establishment of a hierarchy between them. This is necessary for the definition of requirements. These are dependent upon the genotype and performance of animals as well as method of feeding and environment.

To take an example, there are many equations that have been proposed for the estimation of the amino acid requirements of laying hens:

(1) mg methionine/bird per day = $0.037 W + 4.5 \Delta W + 5.39 E$
(2) mg lysine/bird per day = $0.04 W + 8.6 \Delta W + 12.6 E$

where:

W = liveweight (g)
ΔW = liveweight change (g/day)
E = egg weight produced (g/day)

These equations may be applied to individual hens but do not take into account variability within a flock. Moreover it has been frequently established that the biological response to an increase in the dietary level of an amino acid, previously limiting, is curvilinear. Biological requirements therefore may not be precisely defined: it is better to speak in terms of an economic optimum that corresponds to the quantity of an essential amino acid which promotes a maximum margin during production.

Knowledge of the value of proteins

Analysis of raw materials allows an indication of their potential value as protein sources but is insufficient in the prediction of their real nutritional value, in particular if they have been processed.

In the absence of any rapid and routine method which specifically measures amino acid availability, it is necessary to adopt a margin of security within nutritional constraints and analytical results. Values given in the tables of composition of raw materials are the average obtained following numerous measurements but do not necessarily correspond to levels of available amino acids within a given sample.

Chapter 4
Mineral nutrition of non-ruminants

That which follows is not intended to be a consideration of mineral requirements of each species (this will be considered alongside other requirements), nor will it describe the metabolism of each element which would be outside the scope of the present work. It is intended only to draw attention to certain general principles that are frequently overlooked in the calculation of dietary levels.

Application of the concept of requirement in the case of minerals

In order to define the dietary level of an essential mineral it is usual to provide answers to two questions: how much and in what form. Given that these two questions are important, a third should often be added: why. It is becoming more and more evident that:

(1) on the one hand different physiological functions may, in the same animal, require different dietary levels;
(2) on the other hand optimum economic performance may not correspond to optimum biological performance.

Typical examples of this may be found in phosphorus nutrition. Thus it is well known that maximum liveweight gain in pigs and chickens is obtained with a dietary level of phosphorus lower than that necessary to provide satisfactory bone strength, itself lower than that which maximizes bone mineralization. In pigs slaughtered at 100 kg, or chickens at 1.35 kg, maximum bone mineralization is never considered. It is equally known that in chickens the dietary level of phosphorus that promotes the best food conversion ratio is considerably higher than that which optimizes growth. In this case it is an economic compromise, based upon the price of phosphate and the compound feed, that must determine whether the level of phosphorus needs to be increased.

In addition it ought to be noted that all recommendations must above all take into account the level of performance of animals; they must equally allow for a large number of factors, in addition to those directly related to the animal, including for example:

(1) the chemical form of the mineral in the source utilized;
(2) interactions between nutrients;
(3) dietary energy level of feeds;

(4) ambient temperature;
(5) stress arising from ill-health or overpopulation.

These factors modify both the mean and range of performance. They justify the use of margins of security between requirements determined during research and practical recommendations but never to the extent of doubling or tripling the former. Besides the economic wastage which would result, it is known that excess levels, even moderately so, of certain minerals may promote ill-effects. From this point of view, the case of selenium is the best known: an excess of calcium may occasionally promote zinc deficiency in growing rabbits and phosphorus deficiency in lactating does; excess magnesium may reduce the utilization of phosphorus in pigs; excess phosphorus reduces egg-shell weight in the hen.

Determination of mineral requirements

Two complementary approaches may be utilized in the estimation of the requirements of a mineral. Both will be illustrated with respect to calcium and phosphorus which are by far the most important.

The first approach, which may be termed 'general', consists of the distribution among groups feeds containing various levels of the element and recording different biological criteria including:

(1) rate of liveweight gain, feed conversion ratio, incidence of leg weakness all in the growing animal;
(2) litter weights, egg numbers and weights, and milk production all in adult females.

If the various criteria do not produce the same level of requirement, that which optimizes that criterion considered to be the most important is selected. This method is usually confined to poultry.

The second approach, termed 'factorial', bases the estimation of net requirements on their evaluation from each physiological function (maintenance, growth, gestation, egg formation, lactation) and, in their application, includes a factor of availability taking into account true as opposed to apparent intestinal absorption (coefficient of true digestibility — CTD). The determination of CTD requires:

(1) the need to separate faeces and urine (which is impossible in poultry routinely);
(2) the need to estimate endogenous faecal excretion with the use of radioactively labelled elements.

It is therefore considered as a method confined in practice to mammals and one that may only be undertaken in the laboratory.

Besides technical difficulties there are other factors which influence the choice between the two approaches for the estimation of requirements. Thus it is easy to estimate the amount of calcium deposited during pregnancy, in milk, or in egg shells and consequently to adopt a factorial approach. On the other hand net mineral requirements for growth are more difficult to define because the optimum composition of growth is not known. In addition it must be emphasized that even in the same animal, CTD may vary with time: it falls from 70 to 55% in pigs with an increase in liveweight from 10 to 50 kg. In laying hens daily intestinal absorption of calcium increases from 40 to 70% at the commencement of egg-shell formation. It is important therefore not to extrapolate CTD determined under one set of conditions to another.

Given these general principles, it is now proposed to consider the provision of dietary levels of each element or group of minerals.

Problems relating to the principal minerals and their sources

Calcium

With the exception of calcium phosphates, which will be considered subsequently, only calcium carbonates are currently employed in animal feeding. For pure limestone deposits the only problems which can hinder their use is the insufficient solubility of some types of marble utilized in granular form in the feeding of hens: their utilization would therefore result in a lowering of performance and an excessive level of excretion in the faeces.

The biological availability of calcium in limestone is most often between 95 and 100% but may occasionally be as low as 90%. Unfortunately there is as yet no rapid *in vitro* test to verify these differences.

The average level of calcium in biological sources (sea shells, egg shells) is given in *Table 93* (raw materials 190–201). Availability in them is generally good. Other sources of calcium (plaster, cement) present too many difficulties associated with other elements or compounds present (aluminium, silica, sulphate).

It has been shown that, in poultry, the presentation of calcium sources in granular form has a positive result in certain cases, particularly in the laying hen.

Phosphorus

The provision of phosphorus always presents more problems than that of calcium, not only as a consequence of the price of raw materials. It must be borne in mind that in practice no system of non-ruminant animal production (with the exception of adults at maintenance — sows, cocks) is possible if based upon plant feeds without the addition of inorganic phosphorus.

The most frequently discussed problem is the utilization of phytate phosphorus in plant raw materials (but not in stems or leaves) which may represent 60–70% of total phosphorus. It is generally accepted that phytate phosphorus is only sparingly utilized by pigs and not at all by poultry; for the latter it is usual therefore to consider only non-phytate phosphorus, which constitutes a third of total phosphorus in seeds, as available.

Without doubt this is an oversimplification. Thus it is now understood that the utilization of phytate phosphorus is variable in both pigs and poultry as a function of the following:

(1) The initial chemical form: phytates of calcium and phytine (a mixture of Ca, Mg and K phytates) are invariably poorly utilized because of low solubility.
(2) The presence and activity of phytases: wheat, barley and rye frequently contain phytases considerably more active than maize, sorghum and oilseeds.

However there remains the problem of intra-varietal variability which may be important, particularly in wheat.
(3) Raw material processing (cooking, pelleting).

Accepting that knowledge is incomplete, mean values for available phosphorus for poultry presented in the tables of composition, take into account all known published results which consider the actual availability of phosphorus, and only assume a value of 30% of total phosphorus in the absence of specific information.

For the feeding of pigs, this concept of availability is considered too speculative for it to be employed. The use of CTD (which reflects the efficiency of absorption and which therefore incorporates the availability of the element for both digestion and absorption) moreover does to an extent resolve the problem. It considers that the absorptive capacity of the animal is frequently the main limiting factor to availability but does not take into account the metabolic utilization of absorbed phosphorus.

A second problem often encountered is the utilization of phosphorus from inorganic phosphates. This question has been discussed frequently and will not be detailed here: *Table 1* provides a summary of current knowledge in this area.

Table 1 Relative biological value of inorganic phosphates (%). Monocalcium phosphate is the reference material

Product	Poultry		Pigs	
	Average	Range	Relative biological availability	True digestibility (%) in adults
Manufactured products:				
Monocalcium phosphate	100		100	60–70
Dicalcium phosphate anhydrous	85	80–87	–	< 60
Dicalcium phosphate dihydrate	93	85–98	90(70–98)	60–65
Mono/dicalcium phosphate*	95	90–100	–	60–70
Tricalcium phosphate	85	80–100	–	50–60
Mono/dipotassium phosphate	98		–	70–80
Mono/disodium phosphate	98	95–100	100	70–80
Trisodium polyphosphate	92		–	60–70
Mono/diammonium phosphate	–		95	70–80
Phosphoric acid	–		100	80
Calcium magnesium sodium phosphate	98		–	60–70
Naturally occurring products:				
Rock phosphate	60	40–80	–	20–50
Defluorinated rock phosphate	85	80–95	90	–
Degelatinized bone meal	85	80–90	80(60–95)	50–55

Meta and pyrophosphates, colloidal phosphates, and aluminium, calcium and ferric phosphates are only marginally available and not used. They are not represented here.
*A mixture of mono and dicalcium phosphates.

One important practical concern is evident in the verification of actual phosphorus availability in phosphates that are not, or are only partially, hydrated. An extremely sensitive biological test may be conducted with young broilers between 7 and 17 days of age: based upon the mineralization of the tibia or digits in response to feed levels of dietary phosphorus, it allows the quantitative assessment of availability in the source tested with reference to that of a control material (monocalcium phosphate, pure monohydrate).

In pigs an understanding of actual digestibility remains the criterion for reference, but it is difficult to evaluate. A simpler method, developed by Guegen, involves measuring solubility in 2% citric acid; this *in vitro* test gives a range of values from 100 to 10% corresponding to a reduction in CTD in pigs from 75 to 20%. It is therefore of considerable interest and may be widened to include mineral complexes provided that they may contain more than 50% of minerals. In practice it is sufficient to utilize those materials with a solubility equal to or higher than 80–85%.

To conclude the discussion of phosphorus sources it may be noted that the levels of phosphorus and calcium in home-produced animal products (on a dry matter basis) are related by the following equations:

meat meal: $P = 0.78\ Ca - 1.08$
fish meal: $P = 0.66\ Ca - 0.86$

Thus knowledge of one element allows calculation of the level of the other.

Magnesium

Magnesium requirements of non-ruminants are provided essentially by traditional diets based on cereals and oilseed meals and it is rarely necessary in practice to have to provide additional supplements. Possible problems are more likely to be encountered with an excessive level of magnesium which promotes diarrhoea and may reduce the utilization of calcium and phosphorus. For this reason calcium magnesium carbonates are not used.

Sodium, potassium, chloride

It is often the case that diets of plant origin utilized by non-ruminants are deficient in sodium but contain an excess of potassium. The first part of the statement is always so and provision of sodium chloride, which also ensures the supply of chloride, is in practice essential for all species. The second part may require qualification: there seem to be situations where biological performance is improved following potassium supplementation. Potassium requirements are higher with increasing levels of dietary protein (in poultry uric acid excretion is dependent upon potassium) and it is known that potassium supplementation lessens the effects of a lysine-arginine imbalance.

As a consequence of the implications of these electrolytes in maintaining the acid-base and water-mineral balance of animals it is better to consider their provision together rather than separately. In the chicken, Mongin and Sauveur have shown that there is an optimum balance between these elements which ought to be applied: in the absence of a deficiency or toxic excess of any one of them, this relationship may be expressed as $Na + K - Cl$, to be maintained between 230 and 300 mmol/kg of feed.

It is quite likely that similar interactions exist with other animals given that the pig is perhaps less sensitive to an excess of chloride than the chicken. It may be noted as a general principle of adjustment that, initially, the level of chloride (promotor of acidosis) must be fixed strictly according to requirements, to be followed possibly by the addition of sodium and potassium in the form of salts containing metabolizable anions which are promotors of alkalosis (for example bicarbonate or organic anion).

The level of chloride in raw materials is frequently more difficult to establish than sodium which may be measured by flame photometry. To avoid the need for direct determinations, chloride content may be estimated as follows (on a dry matter basis):

maize, wheat, barley, oats: $Cl = 11.6\ Na + 0.034$
meat meal: $Cl = 0.886\ Na - 0.107$
fish meal: $Cl = 1.334\ Na - 0.120$

Trace elements

Although the supplementation of diets with trace elements has become standard practice, situations of deficiency with any one of them may still be apparent. This may be a consequence of:

(1) the estimation of requirements themselves. Methods of determination of trace element requirements are frequently general and as with CTD are too imprecise (in particular with respect to endogenous faecal losses);
(2) considerable variations in raw material composition (see *Table 94*) difficult to estimate in practice;
(3) numerous interactions between trace elements and vitamins etc.;
(4) unavailability.

This latter point is without doubt the most important and makes the exact provision of trace elements with biological activity difficult if the diet is not supplemented. Research into this subject is only fragmentary and concerned mainly with selenium and zinc. For the former, availability is lower in animal meals (only 15–20%) than in plant seeds although the situation is the reverse for zinc (a consequence of complexes with phytic acid in cereal grains).

Those compounds involved in trace element supplementation are given, along with their composition, in *Table 95*, and *Table 96* provides information relating to the relative value of minerals and their oxides. A summary of their use for all species is provided as follows:

Iron Anhydrous or hydrated sulphates, ammonium ferric citrates, choline ferric citrates, ferric chlorides or fumarate or ferrous gluconate are the better sources. Oxides and carbonates are not advised.

Copper All sources (sulphate, carbonate) are useful although oxide, iodide and pyrophosphate are probably less so.

Zinc All simple salts may be used; among ores, only franklinite (a mixture of Fe, Mn and Zn oxides) and sphalerite (sulphide) must not be employed.

Manganese The majority of salts and ores are acceptable with the exception of carbonate ores (rhodochrosite and rhodonite) and certain manganic oxides.

Cobalt	Carbonate, chloride, oxide and sulphate are satisfactory.
Selenium	Selenite is the only form currently utilized.
Iodine	The problem with sources of iodine is instability rather than availability. Stability of iodates (Na, K or Ca) is greater than iodides.
Molybdenum	Sodium molybdate or oxides may be used. Sulphides are not recommended.

Supplementation of trace elements and maximum permissible levels in certain raw materials or compound feeds are subject to official control, as outlined in Table 2.

Table 2 Toxic elements and maximum permissible levels (mg/kg at a moisture content of 12%; (UK Feedstuff Regulations, 1986)

Element	Maximum level		
	Compound feeds	Phosphates	Various raw materials
Arsenic	2	10	Grass meal, dried Lucerne meal, dried clover meal, dried sugar beet pulp, dried molassed sugar beet pulp 4 Fish meal and other marine by–products 10 Other raw materials 2
Fluorine	150 With the exception of feeds for: Pigs 100 Young poultry 250 Other poultry 350	2000	Animal by–products 500 Other products 150
Mercury	0.1	–	Fish meal and other marine by-products 0.5 Other products 0.1
Nitrites	15 expressed as $NaNO_2$	–	Fish meal expressed as $NaNO_3$ 60
Lead	5	30	Grass meal, Lucerne meal, clover meal 40 Yeast 5 Other products 10

For other elements, the maximum permissible levels in compound feeds are Cu: pigs up to 4 months of age 175, older pigs 100, sows 35, poultry 35; Fe: 1250; I: 40; Co: 10; Mn: 250; Mo: 2.5; Se: 0.5; Zn: 250 (mg/kg).

Chapter 5
Vitamin nutrition of non-ruminants

Vitamins are those organic substances whose nutritional role may be defined in two ways:
(1) They are essential, if not for life then at least for the maintenance of normal functions; they cannot be synthesized and thus the animal must obtain them from the diet. In the case of deficiency, health will be adversely affected.
(2) They are active at very low levels: the amounts required in the diet are of the order of mg or µg.

It is not proposed to expand this definition by considering partial synthesis either from provitamins (vitamins A and D) or indirectly from the activity of intestinal microflora (vitamins K and B) as well as conditions where requirements are relatively high (choline). On the whole the characteristics detailed above are those employed in the definition of vitamins.

Characteristics and roles of vitamins

Vitamins control a large number of vital processes. Some outline of their mode of action and the ill-effects arising from deficiencies will be provided. More detailed descriptions may be found in all technical documents. In any case it must be understood that symptoms are frequently not specific and that clinical observations are only at best rough guides.

Table 3 presents a classification of vitamins and their common commercial forms. In the table, as well as the text which follows, the traditional means of dividing vitamins into two large groups based upon their solubility in different solvents is adopted.

Fat-soluble vitamins

These vitamins, which are soluble in fats and their solvents, are deposited in adipose tissue and the liver in relatively significant quantities: accordingly they may be provided on an irregular basis. Alongside this capacity for accumulation, fat-soluble vitamins have a potential toxicity and may adversely influence the health of the animal as well as its performance if they are present in very high levels. Fat-soluble vitamins also have in common a frequent effect at the level of cellular membranes.

Table 3 Classification of major compounds with vitamin activity

Major naturally occurring forms	Commercial forms	Activity expressed as
I. Fat–soluble vitamins		
Vitamin A: A_1 = retinol (alcohol) retinal (aldehyde) A_2 = 3–dehydroretinol	Acetate and palmitate vitamin A_1 L Retinol protected (microencapsulated) or water–soluble	iu, being 0.33 µg of A_1 and, at best, 0.6 µg of β–carotene
Provitamin A: β- and α–carotenes cryptoxanthine	β–carotene	
Vitamin D: D_2 = ergocalciferol D_3 = cholecalciferol	Vitamin D_2 Vitamin D_3, either free or contained within microcapsules	iu, being 0.025 µg of D_2 (pigs, rabbits) or of D_3 (all species)
Vitamin E: α–tocopherol β–tocopherol	*dl* α–tocopherol *dl* α–tocopherol acetate	iu and mg(1 iu = 1 mg of *dl* α–tocopherol)
Vitamin K: K_1 = phylloquinone K_2 = menaquinone	K_1 K_3 = menadione	mg of K_1 and K_3 (1 mg of K_3 = 3.8 mg of K_1)
II. Water–soluble vitamins		
Vitamin B_1: Thiamine (aneurine)	Thiamine hydrochloride Thiamine mononitrate	mg of thiamine hydrochloride
Vitamin B_2: Riboflavin	Riboflavin Na riboflavin 5–phosphate	mg of riboflavin
Pantothenic acid: Coenzyme A	Ca and Na pantothenates Panthenol	mg of pantothenic acid
Vitamin B_6: Pyridoxine Pyridoxal Pyridoxamine	Pyridoxine hydrochloride	mg of pyridoxine hydrochloride
Vitamin B_{12}: Cyanocobalamin Methylcobalamin	Cyanocobalamin Hydroxcobalamin	mg of cyanocobalamin
Niacin (PP): Nicotinic acid Nicotinamide	Nicotinic acid Nicotinamide	mg of niacin
Folic acid: Pteroylmonoglutamate Pteroylpolyglutamates	Folic acid (pteroylmonoglutamic)	mg of folic acid
Biotin: *d*–biotin, free and bound	*d*–biotin	mg of *d*-biotin
Choline: Choline (esterified)	Choline chloride	mg of choline (1 mg = 1.15 mg of chloride)
Vitamin C: Ascorbic acid Dehydroascorbic acid	Ascorbic acid Calcium ascorbate Sodium ascorbate	mg of *l*-ascorbic acid (1 mg = 1.13 mg of ascorbate)

Vitamin A influences vision, steroid synthesis (sexual and adrenal hormones), and the formation and maintenance of epithelial cells. Deficiency, particularly serious during growth, is damaging at any age. Symptoms, which affect all those functions mentioned, have a relatively specific nature in the eye (blindness) and the skin (hyperkeratosis or scurf in the pig). The deficient animal is more prone to infection, and mortality is considerably increased.

Included in vitamin A are a number of other molecules with varying activity: A_1, retinol and retinaldehyde have identical maximum activity whereas retinoic acid, A_2 or dehydroretinol and dehydroretinaldehyde only have a partial activity. Certain carotenoids, including in particular β-carotene, may be metabolized into vitamin A by the organism and constitute, by virtue of this, provitamins. (1 iu of vitamin A = 0.33 μg of A_1 and, at best, 0.6 μg of β-carotene.)

Vitamin D is essential for the absorption of calcium and subsequent bone calcification. Deficiency promotes rickets in the young with limb deformities and swelling of joints (crooked breastbone in poultry). In adults, bones demineralize and become fragile (osteomalacia). Recent studies have shown that vitamin D_3 (cholecalciferol) is only a prohormone activated by hydroxylation in the liver and kidneys. The resultant molecule (1,25-dihydroxycholecalciferol) acts as a true hormone in a number of target organs (intestine, bone, kidney, parathyroid gland); its synthesis is controlled by feedback mechanisms (negative) and to a greater or lesser extent by calcium absorption.

Vitamin D_2 (ergocalciferol) has no activity in birds. In all species dietary provision of vitamin D is complemented by synthesis in the skin if the animal is exposed to ultraviolet rays from the sun. (1 iu of vitamin D = 0.025 μg of D_2 or D_3.)

Vitamin E, of which the most active form is α-tocopherol, is concentrated in the membranes of organelles or cells as a consequence of its affinity both for lipids (through its side chain) and proteins (through its tocopherol moiety). Thus attached it would play an antioxidant role in specifically protecting membranes in reactions that involve sulphur molecules (glutathione) and selenium. At high pharmacological levels (20 times the requirement) vitamin E acts as a stimulant of immune responses (through synergism with selenium).

Vitamin E deficiency will promote variable effects dependent upon species, including a slower growth rate which is particularly evident in muscular tissue (muscular dystrophy observed in pigs, rabbits and poultry alike) and mulberry heart disease in pigs. Hepatic necroses are observed only in pigs, and exudative diathesis and encephalomalacia only in poultry. Although originally considered as a vitamin influencing fertility, α-tocopherol does not appear to be important in this respect except in the rat.

There are many tocopherols and tocotrienols with very variable vitamin E activity. Vitamin E activity is expressed in terms of mg dl α-tocopherol acetate (1 mg = 1 iu) although the most active form is d-α-tocopherol (1.5 times more active than the racemic form).

Vitamin K is provided in higher animals by two complementary routes: via the diet and via an important contribution from synthesis by intestinal microflora. Deficiencies only appear under certain circumstances when microfloral activity is inhibited (following sulphamide use) when coprophagy is prevented (of probable importance only in the chicken) or in the presence of antivitamin factors (natural-mouldy forage, or artificial-warfarin). As a consequence of an accident during rearing or deliberately as a result of an experiment, a deficiency in vitamin K is associated with a considerable reduction in blood prothrombin levels. This is followed by an important increase in coagulation times; even minor damage will result in serious internal haemorrhage and mortality may be extremely high.

Many compounds have vitamin K activity: vitamin K_1 or phylloquinone of green vegetables, vitamin K_2 or menaquinone synthesized by microflora, and vitamin K_3 or menadione or menaphthone which is only a provitamin transformed into K_2 in the animal by the addition of a side chain. (1 mg of K_3 is equivalent to 3.8 mg of K_1.)

Water-soluble vitamins

Besides their solubility in water, vitamins of the B group have, in common, a function as enzyme cofactors and, by virtue of this, a key role in the anabolic and catabolic processes involved in the renewal of body tissue.

They are not only found in higher animals but also in microorganisms. The latter may occasionally be able to synthesize them whereas the former must obtain them from natural sources to ensure their survival.

In contrast to the situation with fat-soluble vitamins, B vitamins do not accumulate in the organism. Therefore a suitable daily amount must be provided during production, otherwise there will be a rapid reduction in performance. In situations of deficiency, the animal will save vitamins to ensure its survival. Concentrations of them in milk and eggs are influenced earlier than the overall level of production such that a slight deficiency can adversely influence the development of the young without ill-effect to the mother. As a consequence of their water-soluble nature, excessive levels of B vitamins are rapidly excreted and, in contrast to fat-soluble vitamins, are not toxic.

Vitamin B_1 or thiamine plays a fundamental role in carbohydrate metabolism (oxidative decarboxylation). The nervous system is an important user of carbohydrates and is particularly sensitive to B_1 deficiency (hence its former name aneurine). Both loss of appetite and a general weakness may be associated additionally with symptoms of nervous disorders in the young chicken and ataxia in pigs. In addition, cardiovascular problems may be observed together with a slower heart rate and oedema in pigs. Vitamin B_1 activity is expressed in mg of thiamine hydrochloride.

Vitamin B_2 or riboflavin is present within those enzymes (flavinoids) involved in hydrogen transport and participates in respiration which assures the use of oxygen by the organism. It influences all tissues and a deficiency of riboflavin is associated above all with non-specific symptoms such as a lowering of appetite and rate of growth. In the young chicken lameness is observed and occasionally a characteristic deformation of the toes which curl inwards (curled toe). In pigs, skin lesions (dry skin) are frequent. Vitamin B_2 activity is expressed in mg of riboflavin.

Pantothenic acid is rarely found in a free form in nature. It is on the other hand an integral constituent of coenzyme A which is responsible for the transfer of the acetyl radical. By virtue of this, it plays a central role in nutrient interconversions: fatty acids, carbohydrates and amino acids. Deficiency symptoms are varied and range from a loss in appetite and weight to alterations in the skin and neurological disorders ('goose-stepping' in pigs). Pantothenic acid is too unstable in its free state and is utilized in the form of a salt (calcium pantothenate) or alcohol (panthenol); 1 mg of pantothenic acid corresponds to 1.09 mg of calcium pantothenate.

Vitamin B_6 or pyridoxine is present in many diverse enzymes which control amino acid metabolism: transamination, decarboxylation and deamination. The role of proteins in all vital processes results in few specific deficiency symptoms: a fall in appetite and performance, changes in the skin and mucosal membranes. Three similar compounds have virtually the same activity: pyridoxine (or pyridoxol), pyridoxal and pyridoxamine. Vitamin B_6 activity is expressed in mg of pyridoxine hydrochloride (equivalent to 0.82 mg of each of the three compounds).

Vitamin B_{12} is involved as a cofactor in nucleic acid synthesis. As a consequence of this, a deficiency particularly restricts cellular division: fetal development, growth and erythrocyte generation. Deficiency symptoms are dependent upon the diet and rearing conditions. Among common raw materials, only those of animal origin contain B_{12} which has been named 'animal protein factor'. Synthesis by intestinal microflora is considerable and a diet based entirely on plant products will only result in deficiency if coprophagy is prevented. There are many forms of B_{12} of identical activity: their complex formulae are the same with the exception of the radical OH, CN or CH_3, which stabilizes the molecule via a complex with cobalt at its centre giving respectively hydroxy, cyano or methyl cobalamin. Activity is expressed in terms of mg of cyanocobalamin (the most stable form).

Niacin is part of the group of enzymes active in hydrogen transport and, like B_2 (considered above) is concerned in respiration which controls oxygen utilization. A deficiency results in a lowering of performance, lesions of the skin (dermatitis) and mucosa, and is invariably responsible for lameness in poultry. There are two forms with the same activity: nicotinic acid (niacin) and the corresponding amide (nicotinamide or niacinamide). Tryptophan, an amino acid present in proteins, is a provitamin which may contribute significantly towards requirements if it is present at high dietary levels. Activity of the vitamin is expressed in mg of niacin.

Folic acid or folacin is essential for the transfer and transformation of single carbon radicals which are fundamental to metabolism. Consequently it is concerned with the biosynthesis of nucleic acids in relationship with B_{12}. A deficiency in either vitamin, moreover, will result in similar symptoms. Lack of folic acid in the diet may result in a reduction in appetite, a lowering of growth rate, reproductive problems and anaemia. In poultry, it may occasionally be associated with feather depigmentation and lameness (chondrodystrophy in young turkeys). The correct term for folic acid, pteroyl monoglutamic acid, is based upon only one glutamic acid moiety in its molecule. In nature, conjugated forms are most often found (pteroyl polyglutamates) where each molecule has the same activity and the same structure except if it contains several glutamic acid moieties (3–7) in its chain. Activity is expressed in mg folic acid (monoglutamic).

Biotin acts as a coenzyme and is concerned specifically with the transport of carboxyl radicals. It is responsible for the carboxylation of acetyl CoA to malonyl CoA which is the first step in the synthesis of fatty acids. It is equally involved in the synthesis of amino acids and purines. Biotin deficiency has been observed particularly in pigs and poultry and is most frequently associated with skin lesions, inflammation of feet, cracks in foot pads and a reduction in fertility. Appetite and rate of growth are reduced. Distortions in the metatarsus may be observed which results in perosis (cf. choline deficiency) in chickens and, above all, in young turkeys. Two optically active isomers exist: *d*- and *l*-biotin. Only *d*-biotin has vitamin activity (expressed in mg) which is identical for both the *d*-α and *d*-β forms.

Choline is involved in transmethylations as a donor of methyl radicals. However it does not act as an enzyme cofactor and therefore has a separate function. Moreover it is difficult to assess its role because it is a widely found constituent present in phospholipids and therefore requirements for it are relatively high. Nevertheless a deficiency will influence performance. In addition to general problems (reduction in appetite and growth rate) specific symptoms include fatty liver (all species) followed by cirrhosis, necrotic kidney which may result in death (rabbits), a specific perosis in birds (chickens, young turkey) which is characterized by a rotational twisting of the tibial-metatarsal joints such that when both limbs are involved any movement is impossible, and spasmodic paraplegia in the hind limbs in piglets.

Choline is synthesized *in vivo* from the amino acid serine, and the necessary methyl groups are supplied from methionine or betaine. During those times when requirements are high (young broiler, pregnant sow) this capacity for synthesis may, however, be insufficient; choline (in its basic form, as its chloride or as a constituent of phospholipids) must therefore be provided as such in the diet. Activity is expressed in mg of choline (equivalent to 1.15 mg of choline chloride).

Vitamin C (ascorbic acid) is too important to be omitted from this list. However it must be understood that poultry, pigs and rabbits can easily synthesize it from glucose and it is therefore not essential for these species. Its use as an additive is more for pharmacological than nutritional reasons. The probable role of vitamin C in the synthesis of corticosteroid hormones may explain its 'anti-stress' effect. Thus it may be of use when there are specific problems of adaptation involving changes in temperature and ill-health associated particularly with weak animals (early-weaned piglets). Vitamin C activity is expressed in mg pure crystals of *l*-ascorbic acid (equivalent to 1.13 mg sodium ascorbate and 1.12 mg calcium ascorbate).

Vitamin requirements and diet formulation

Experiments to determine requirements, using small numbers of animals, are generally based on semi-synthetic diets supplemented with purified or synthetic vitamins. Values obtained are low and are presented in *Table 4* as guidelines. These are in effect the minimum levels which ought to be maintained in formulation. In practice, the levels in diets must be between 2 and 5 times higher to promote maximum levels of performance. Many variable factors account for this: some associated with the animal and others with the diet consumed.

Among a group, individual animals have differing vitamin requirements; dietary levels must satisfy those that are the most needy. Moreover the determination of minimum requirements is very difficult and the absence of deficiency symptoms is not proof that levels are adequate. On the other hand, it must be understood that feed composition influences requirements; one high in carbohydrates increases the requirement for B_1. Without entering into complex details of the reactions involved, it is sufficient to state that the concept of requirement is relative and is based upon rearing conditions (temperature) and, above all, health status of the herd. Thus vitamin A requirements are increased considerably by infections or the presence of parasites.

The contribution of raw materials to the provision of vitamins is also difficult to determine precisely; their origin, processing and methods of storage have a considerable influence on their vitamin levels. Figures for composition given in *Table 97* (vitamin levels of raw materials) are only mean values. Levels are also dependent upon the methodology employed.

Finally it is not sufficient for the vitamin to be well preserved in the feed; it must also be active. This raises the problem of antivitamins, substances which destroy or complex vitamins, or those which tend to replace vitamins in biological reactions but do not have their activity.

Fat-soluble vitamins

In feed formulation vitamin A (retinol) and vitamin D (D_2 only for pigs and rabbits, D_3 for all species) are usually added to diets for non-ruminants without taking into account their levels in raw materials. Recommended levels (detailed subsequently for each species) which are between 4 and 8 times higher than those suggested as minimum requirements, ensure a sufficient margin of security. Instability frequently associated with vitamins A and D (A is very susceptible to oxidation) is a very important factor in the choice of the commercial form of the vitamin (encapsulation and stabilization in microparticles). With the products used, conservation in compound feeds is excellent.

Taking into account the possibility of variable requirements (health status) it would be dangerous to reduce the recommended levels of vitamins A and D for no good reason, bearing in mind the very low price of commercial preparations. On the other hand it must be understood that vitamins A and D may have a toxic effect: a reduction in performance with excess consumption of vitamin A and calcification away from bones with vitamin D. In practice, dietary levels of vitamin A of 20 000 iu/kg and 2000 iu/kg for vitamin D should never be exceeded.

In contrast to the other fat-soluble vitamins, vitamin E (dl-α-tocopherol acetate) is commercially expensive. This explains why it is used sparingly and why, as a consequence, deficiency symptoms are found. It must be said that there are an increasing number of problems associated with vitamin E and requirements are based upon a number of factors (genetic and nutritional) which are poorly understood. The levels presented in the tables of raw material composition are only mean values which do not reflect the variability in analytical results. Levels may be considerably reduced by bad processing or poor storage, as tocopherols are easily destroyed by oxidation. It would seem imperative therefore that dietary levels should be higher than the minimum requirements (at least double the values given in *Table 4*). At the same time, those factors which contribute to higher requirements should be avoided by ensuring that there are no dietary deficiencies in selenium and sulphur amino acids and by using antioxidants at the legally authorized levels, particularly in diets high in fat.

Supplementary levels of tocopherol are important not only for performance during rearing but also for the quality of the carcass. In pigs, the occurrence of pale meat (other than the 'pale soft exudative' condition) may be reduced with vitamin E. The development of rancidity during cold storage is also reduced.

The stimulation of immune responses is only achieved with very high dietary vitamin E levels (150–300 mg/kg) which may be employed on a therapeutic basis or as a preventative measure during periods of risk. There are hydro-dispersible preparations of vitamin E which allow it to be added to water for short periods of treatment.

By virtue of intestinal synthesis, the need for dietary levels of vitamin K is only really necessary in rabbits and pigs during the first few weeks of life. On the other hand, it is important to add vitamin K to all diets for poultry due to the short length of the digestive tract in this species and their rearing conditions which reduce coprophagy. The recommendations given for each system of production account for an increase in requirements in those suffering from coccidiosis. They are not on the other hand sufficient to counteract a deficiency caused by the presence of the antivitamin K factor dicoumarol in mouldy feed.

Water-soluble vitamins

In contrast to ruminants, non-ruminants derive very little benefit from B vitamins synthesized by intestinal microflora. The localization of the microflora towards the end of the digestive tract reduces the recovery to low levels during passage of digesta, and modern systems of rearing diminish the possibilities of coprohagy in poultry and pigs. Only the rabbit is able to benefit from synthesis by intestinal microflora by directly recovering the soft faeces (pellets voided during the night) from the anus. Dietary levels of B vitamins are essential for pigs and poultry whereas for rabbits they provide at the best a precautionary amount for intensive production systems.

Under normal conditions of feeding diets based on cereals or their by-products, thiamine (B_1) requirements are largely met. A deficiency is however possible under some circumstances where cereals are replaced by other raw materials including potatoes, cassava and proteins low in vitamins (gluten, animal meals) or those with levels that have been reduced due to excessive heat treatment.

Vitamin B_2 is abundant in milk products, certain plant materials (lucerne) and animal meals but is found only at low levels in seeds and, consequently, it is important to include it in typical diets based on cereals and oil meals. Riboflavin is unstable only in the presence of light and is well preserved in raw materials and compound feeds. A relative degree of confidence may therefore be placed in the tables of composition, and supplementation need only be as a precautionary measure, as the vitamin is inexpensive, to give a dietary level of the order of double the minimum requirement.

As its name suggests, pantothenic acid is widespread among all raw materials. In these, as well as in commercial forms (pantothenates), it is stable and preserves well. As with riboflavin, it is possible therefore to calculate dietary levels from tables (*see Table 97*) and to provide a precautionary amount double the theoretical requirement to cover variability in needs.

The measurement of vitamin B_6 is difficult because allowances must be made for the three active forms (pyridoxine, pyridoxal, pyridoxamine) to produce a meaningful figure. The methodology employed for evaluation of raw materials varies from one centre to another and largely explains the variability of results. In addition, the vitamin is, to an extent, unstable in the presence of heat, light and pro-oxidants. As a consequence of these conditions, the mean values provided in the tables only allow a rough estimate of levels in raw materials. Antivitamin B_6 factors, present in linseeds, are fortunately not found in the more common raw materials. There is ultimately little risk of deficiency and a precautionary supplement (pyridoxine hydrochloride which is stable in compound feeds) may be considered, particularly early on in life.

Vitamin B_{12} requirements are extremely low. In rabbits they are met by coprophagy and, in other non-ruminants, by small amounts of animal meals. When the diet is entirely of plant origin, cyanocobalamin must be added to meet requirements but unnecessarily high levels are to be avoided.

Niacin is without doubt the most stable of the B vitamins. Values provided for the different raw materials are comparable between centres and may therefore be used in the calculation of dietary levels. It must be understood however that the vitamin may be partially unavailable (maize) and that the existence of antivitamin factors has been pointed out. Certain common feed formulations (maize:soya) being relatively low in niacin may promote a deficiency despite the capacity of those species considered in this book for the synthesis of niacin from tryptophan. This is particularly so for the turkey poult (*Table 4*) but also for young web-footed birds and the guinea-fowl. A precautionary supplement is necessary if the dietary level of niacin calculated from raw materials is lower than 1.5 times the requirements of the animal.

Table 4 Approximate minimum vitamin requirements of pigs, poultry and rabbits (in iu or mg/kg feed)

	Pigs			Poultry			Young rabbits
	Young pig	Fattening pig	Breeding sow	Young broiler	Young turkey	Laying hen	
Fat-soluble:							
Vitamin A(iu)	2200	1300	2000	1500	4000	4000	2000
Vitamin D(iu)	220	150	200	200	900	500	350
Vitamin E	11	11	10	10	10		5
Vitamin K	0.07			0.53	0.7		1
Water-soluble:							
Vitamin B_1	1.3	1.1	1.0	1.8	2.0	0.8	4
Vitamin B_2	3.0	2.2	3.0	3.6	3.6	3.8	6
Ca pantothenate	13	11	12	10	11	10	20
Vitamin B_6	1.5	1.1	1.0	3	4	4.5	1.0
Vitamin B_{12}	0.022	0.011	0.015	0.009	0.003	0.003	0.015
Niacin	22	12	10	27	70	10	15
Folic acid	0.6	0.6	0.6	0.6	0.9	0.35	
Biotin	0.1(?)	0.1(?)	0.1(?)	0.09	0.03	0.15	
Choline	1100(?)	600(?)	1250(?)	1300	1900		1300

It must be emphasized that these figures are only guidelines (*see* text); additions to diets must be liberally estimated. Precise recommendations will be found within those chapters dealing with the production of individual species.
Maximum permissible levels in complete feed (in iu/kg) are vitamin D: pigs 2000, piglet milk replacer feeds 10 000; vitamin D_3: chickens and turkeys for fattening 5000, other poultry 3000.

Folic acid requirements are low because of the contribution of synthesis by intestinal microflora. Dietary levels are not essential except with poultry (reduced role of microflora, variability in requirements) and pigs during certain critical periods (very young, pregnant). The methodology for determining levels in raw materials does not always take into account the conjugated form (around 80% of the total) and some have underestimated dietary levels. In practice, diets based on soya-bean meal (particularly rich) prevent a deficiency. On the other hand, those based on protein meals which have considerably lower values (lupin) and associated with cereals (or, even worse, cassava) may promote one. The possibility of a deficiency is increased in the presence of antivitamin factors (some synthetic forms which moreover are employed as coccidiostats in rabbits and poultry) or following treatment with sulphamides. The possible level of supplementation in the form of folate must take into account all these factors and also the instability of the vitamin (losses of up to 50% may be possible following several months' storage). All things considered, it would seem prudent to meet requirements of those species which rely upon levels in the diet (poultry) by providing at least 2–3 times the dietary amounts indicated in *Table 4*.

The presence of biotin in virtually all raw materials and the small amount required (except in turkey poults) ought, more or less, to remove all risks of deficiency. The inability of certain diets (wheat:meat meal) to meet requirements may be demonstrated in the chicken as well as the pig (lesions in the hoofs). Among the factors responsible, variability in the level in cereals based upon harvesting conditions and variety is certainly important. In addition, the biotin levels do not appear to be associated with vitamin activity. Recent studies in the chicken have shown that a significant amount of the vitamin (or all of it in the case of wheat) is present in an unavailable form: the values given in *Table 5* apply to the chicken but it is quite likely that they are, more or less, equally applicable to other species (pigs). Biotin requirements may be increased by moulds which secrete an antivitamin factor (streptavidine) that inactivates biotin by forming a stable complex. Supplemental dietary levels of biotin would therefore appear to be justified. There will however be problems associated with the high price of this vitamin, and both the raw materials employed together with the importance of requirements must be taken into account. Conversely, supplementation must be considered essential for turkey poults.

Table 5 Biotin availability in raw materials (%), after Frig (1976, 1977)

Cereals		Protein feeds	
Oats	30	Wheatgerm	40
Wheat	0	Maize gluten	100
Maize	100	Lucerne	100
Barley	0–20	Peanut meal	54
Sorghum	20	Rapeseed meal	76
		Soya-bean meal	100
		Brewer's yeast	100
		Fish meal	100
		Meat meal	100

The ability of animals to synthesize choline is insufficient to cover requirements under intensive conditions even if the diet is well supplied with methionine. This is particularly true with rabbits and poultry during the first few weeks of life and in the pregnant sow. Under these conditions a precautionary supplement must be provided to give diets containing at least 1.5 times the amount indicated in *Table 4*. Commercial forms available (based on choline chloride which is very hygroscopic) supply 25–50% (dry product) or 70% (liquid product) of choline.

Vitamin C is not essential for poultry, pigs or rabbits. Moreover it must be understood that it is very unstable and that there is actually no active stable form at present: direct supplementation of compound feeds is therefore of no interest because the vitamin would be rapidly destroyed (oxidized irreversibly) following contact with minerals. Only concentrated mixtures (anti-stress feeds) or the vitamin itself may be used on an *ad hoc* basis.

Conclusion

No one vitamin is in itself more important than another. The only one that is important is the one that is deficient in a diet as it would constitute a limiting factor in the utilization of that diet. Therefore it is important to supply all vitamins, even the most costly. (As many vitamin deficiencies observed are associated with tocopherol and biotin, the cause may be associated with their high price.) On the other hand, it is not worth raising levels of water-soluble vitamins above usual recommended levels, and it may be harmful to do so with fat-soluble vitamins. Recommendations provided under each production system carry with them a margin of security sufficient for common diets. Only with adverse rearing conditions (stress, ill-health) or with the use of new raw materials may departures from these norms be exceptionally justified. In no case does the accumulation in one vitamin allow the saving of another. Therefore for each vitamin taken in isolation, two dangers must be avoided: wastage and under-provision, remembering that the second is by far the more important.

Finally, it ought to be remembered that measurements of vitamin levels frequently do not allow for factors including their storage and availability nor the presence of antivitamin factors which determine how effective they are in the diet.

Part II
Dietary recommendations

Chapter 6
Nutrition of growing pigs

Synopsis and mean recommendations

Whatever its stage of life, recommended nutrient allowances for the pig may be expressed in terms of daily amounts as a percentage of the feed or as a ratio to energy. Most frequently they have been obtained during trials in animals with the aim of simultaneously optimizing several measurements of production (growth rate, feed conversion ratio, carcass quality). It is equally possible in certain cases to resort to a factorial approach in the determination of requirements which simultaneously takes into account the nature and quantitative importance of needs (maintenance, body tissues) as well as the efficiency of utilization of nutrients corresponding to these needs. This method is however not currently used except for minerals, which are the only nutrients for which the components of requirements and efficiencies of utilization are sufficiently well known. It would be interesting in the future to be able to utilize the factorial method for other nutrients as this would allow variations in requirements based on performance to be precisely taken into account.

The rearing period of pigs destined for consumption may be divided into four phases which are based on both liveweight and the buildings used:

(1) Suckling phase (pre-weaner) corresponding to ages between 21 and 40 days (21–28 days in the case of early weaning). Liveweight increases from 5 to 10 kg.
(2) Post-weaning phase (weaner) from 40 days (or 28 days for early-weaned pigs) until entry into the fattening house. Liveweight increases from 10 to 25 kg.
(3) Growing phase from 25 to 60 kg liveweight.
(4) Finishing phase from 60 kg liveweight until slaughter, frequently at 100 kg.

Prior to considering each nutrient in detail, mean recommendations for energy, protein, amino acids and macrominerals for the growing pig are presented in *Table 6*, and *Table 7* indicates recommended levels of trace elements and quantities of vitamin additions. In order to fulfil the needs of feed formulation, these recommendations are expressed as a concentration (energy) or as a percentage (protein, amino acids, minerals) of the feed of known energy value.

Table 6 Recommended dietary levels of energy, protein, amino acids and minerals for pigs (on an as fed basis)

	Stage of growth			
	Suckling	Weaner	Grower	Finisher
Liveweight range (kg)	5–10	10–25	25–60	60–100
Age range (d)	21–40	40–70	70–130	130–180
Dry matter (%)	90	90	87	87
Energy concentration (kcal DE/kg):				
Range*	3300–3600	3300–3600	3000–3400	3000–3400
Mean*	3500	3500	3200	3200
Crude protein (%):				
Usual quantity	22	19	17	15
Minimal amount of ideal protein**	20	18	15	13
Amino acids (%):				
Lysine	1.40	1.20	0.80	0.70
Methionine + cystine	0.80	0.70	0.50	0.42
Tryptophan	0.25	0.22	0.15	0.13
Threonine	0.80	0.70	0.50	0.42
Leucine	1.00	0.87	0.60	0.50
Isoleucine	0.80	0.70	0.50	0.42
Valine	0.90	0.76	0.55	0.50
Histidine	0.34	0.32	0.20	0.18
Arginine	0.36	0.35	0.25	0.20
Phenylalanine + tyrosine	1.30	1.10	0.80	0.70
Minerals:				
Calcium	1.30	1.05	0.95	0.85
Phosphorus	0.90	0.75	0.60	0.50

*Or, using ME (where ME=0.95 DE) a mean concentration of 3320 kcal/kg in diets for suckling/weaner pigs (range of 3130–3420) and 3040 in diets for growing/finishing pigs (range of 2850–3230).
**After supplementation with free lysine: 55% lysine in protein.

Energy nutrition

Energy levels of diets for pigs are generally expressed in terms of digestible energy (DE — see Chapter 2; if metabolizable energy (ME) is used, DE values may be multiplied by 0.95 to convert them). For the needs of feed formulation, energy levels are invariably expressed in terms of a concentration (kcal DE/kg feed). This recommendation is in fact the only one that operates in the case of the animal fed *ad libitum*.

If a restricted feeding system is used, then it is the daily amount of feed, of a known energy value, that is most often required (kcal DE/animal per day) as a function of age (or, more frequently, of liveweight). In this case, expected performances (daily liveweight gain, feed conversion ratio, carcass characteristics) for a given type of pig (based on sex and genotype) are determined by the degree of feed restriction selected and by the climatic conditions operating.

Table 7 Recommended added dietary levels of trace elements and vitamins for the growing pig (in iu/kg or mg/kg)

	Young pig	Growing pig
Trace elements (mg):		
Iron	100	80
Copper	10	10
Zinc	100	100
Manganese	40	40
Cobalt	0.1–0.5	0.1
Selemium	0.3	0.1
Iodine	0.6	0.2
Fat-soluble vitamins:		
Vitamin A (iu)	10 000	5 000
Vitamin D (iu)	2 000	1 000
Vitamin E (mg)	20	10
Vitamin K (mg)	1	0.5
Water-soluble vitamins (mg):		
Thiamine	1	1
Riboflavin	4	3
Calcium pantothenate	10	8
Niacin	15	10
Biotin	0.1	0.05
Folic acid	0.5	0.5
Vitamin B_{12}	0.03	0.02
Choline chloride	800	500

Energy concentration

Ad libitum feeding

The pig at weaning For the suckling piglet (between 5 and 10 kg liveweight) the constraints associated with the introduction or exclusion of certain raw materials are more important than the provision of a mean energy level in feed formulation. It would be useful, for example, to be able to maintain 10% of milk products in the diet and it is not advised to add more than 5% of fat; in addition there should be a minimum level of between 3 and 4.5% of crude fibre. It is possible to conclude, therefore, that the optimum energy concentration of the diet for piglets of this age should be in the region of 3300–3600 kcal DE/kg (the values given in the tables) which will ensure, right from the start, successful development of the digestive tract.

It is known that if measurements of feed utilization are carried out with similar piglets recently weaned (5–10 kg liveweight), values obtained will be between 5 and 10% lower than those calculated. These differences are associated with problems of adaptation of the young animal to its feed. They are largest with fats, slightly less with protein sources and minimal with starchy feeds low in fibre, above all if the compound feed is pelleted.

During the second phase (10–25 kg liveweight) the piglet responds to variations in energy concentration in the same way as the growing/finishing pig.

The growing/finishing pig   The energy level of a feed may vary within a relatively large zone associated with the ability of the pig to maintain a level of energy intake. The extent of this adjustment varies however with the energy concentration of the feed; thus with a dilution in dietary energy level (within the common range of 2900–3400 kcal DE/kg) the growth rate relative to the daily amount of feed consumed is proportionally less than the reduction in dietary energy level. The result of this is a small reduction in total energy intake associated with a slight fall in rate of liveweight gain but, above all, an appreciable decline in fat content of carcasses at slaughter. Energy conversion efficiency (kcal DE/kg liveweight gain) is hardly affected.

With animals that have a strong potential for fat deposition it may be necessary to limit the energy level of the feed to the range 3000–3200 kcal DE/kg in order to obtain satisfactory carcass quality. On the other hand, those pigs with a high potential for muscle development (e.g. Pietrain) may be fed *ad libitum* with diets of a high energy concentration (up to 3400 kcal DE/kg) with no resultant excessive carcass fat content. This difference is based upon the fact that the tendency of the pig to overconsume is less important when its rate of fattening is not so pronounced, above all with diets rich in energy.

In addition, variations in dietary energy concentration may also be influenced by constraints imposed by preferential utilization of certain raw materials: thus a limit of 3000–3100 kcal DE/kg is imposed for diets based on barley against 3300–3400 kcal DE/kg for those based on maize.

Table 8 presents a summary of the above discussion on the energy concentration of diets when fed *ad libitum*.

Table 8 Range of energy levels possible in diets for growing pigs

Stage of growth	*Dietary energy level* (kcal/kg)
Young pig:	
Suckling (5–10 kg)	3300–3600
Weaner (10–25 kg)	3300–3600
Growing/finishing pig (25–100 kg):	
Lean type	3000–3400
Conventional type	3000–3200

Restricted feeding

Whatever the stage of production considered the choice of energy concentration determines, at the time of imposition of restriction, the amount of compound feed to be provided. The need for better feed conversion ratios may in this case favour those diets that are relatively rich in energy.

Usual daily allowances of energy-feeding systems

Expected levels of performance following the introduction of a system of feeding are given in *Table 9*.

Table 9 Expected levels of performance in pigs given the recommended feed allowances

Young pig*	Suckling	Weaner
Age (d)	26–40	40–70
Daily liveweight gain (g)	200–250	500–550
Feed intake during period (kg)	5.5**	25.0
Food conversion ratio	1.40	1.65
Growing/finishing pig	Gilts	Castrates
Daily liveweight gain (g)	700–750	650–700
Daily feed intake (kg)	2.1–2.2	2.0–2.1
Food conversion ratio	3.0–3.2	3.2–3.4
Lean: fat ratio	2.4–2.6	2.0–2.3
Lean meat content of carcass (%)	50–52	48–50

*Average age of castration = 26 days.
**Including 450 g of feed consumed prior to castration.

Systems of feeding piglets

Suckling piglets (5–10 kg liveweight) Whenever it is possible without health risks, piglets are fed *ad libitum* after weaning. Their average intake is very variable and is a function of many factors including weaning age, pre-weaning feed consumption, state of body reserves, liveweight.

If health risks associated with digestion are a problem, it may be possible to adopt a regular progressive increase in feed consumed during the two to three weeks following weaning. Such a scheme, together with its format, is given in *Table 10*.

Table 10 Daily allowances for young pigs

Age (d)	Liveweight (kg)	Food intake* (g/d)	Energy (kcal DE/d)
Suckling			
22	5.7	85	300
25	5.8	130	455
28	6.0	180	630
32	6.8	270	945
35	7.6	330	1155
39**	8.8	440	1540
42**	10.0	500	1750
Weaner			
	10	500	1750
	15	770	2700
	20	1050	3600
	25	1200	4200

*Dry matter content of 90% and energy level of 3500 kcal DE/kg.
**Gradual introduction of weaner diet.

Post-weaning piglets (10–25 kg liveweight)   The use of early weaning (around 26 days of age) is followed by a rapid growth rate if piglets are fed *ad libitum*. This is a system used commercially in the production of a heavier piglet (25–30 kg) but one that is also fatter and with a delayed entry into the growth period (i.e. growing/finishing). The adoption of such a scheme for feeding principally at the end of the post-weaning period, using the amounts indicated in *Table 10*, may be justified under these conditions.

Scheme of feeding during growing/finishing

Generally the consumption of energy by a pig fed *ad libitum* is between 3 and 4 times the maintenance requirement (which is itself proportional to metabolic liveweight, at 105–115 kcal $DE/W^{0.75}$), and is in the range 300–400 kcal $DE/W^{0.75}$. This energy consumption for a given liveweight and daily liveweight gain varies considerably according to the relative development of lean and fat tissue. In particular it is approximately 6–8% higher in castrates than in both females and entire males, which have more or less the same level of intake as each other.

With those pigs commonly reared in France the objective of achieving carcasses of lower fat content generally implies some form of feed restriction during the growing/finishing period. Therefore there needs to be a compromise between the various performance criteria of growth rate, feed conversion ratio and the proportion of lean in the carcass.

The adoption of this scale of feeding must take into account differences in tissue growth as a consequence of sex, the effect of castration and genotype. The average energy content of liveweight gain therefore varies, according to animals, to between 3000 and 4000 kcal/kg (with a liveweight increase from 25 to 100 kg). More precisely, an increase in the potential for muscle development, at a given liveweight, is associated with a reduction in the mean energy costs of liveweight gain and, as a consequence, in energy requirements. At an equal liveweight, therefore, leaner animals have a lower energy requirement. As a corollary, for the same level of feed intake based upon liveweight, lean pigs will have a better gain in weight and of lean tissue than fat pigs; simultaneously there would be an improvement in feed conversion ratio. This is the case, for example, with both entire males and females when compared with castrates.

An appreciation of these points ultimately allows the use of a progressive degree of feed restriction from an early age and a differentiation between sex in terms of feed allowances — being liberal in females (90–95% of *ad libitum*) and relatively lower in castrate males (75% of *ad libitum*) during the finisher phase.

In practice, the adoption of a scheme of feeding equally must take into account economic factors. The choice between *ad libitum* and restricted feed intake in fattening pigs integrates such criteria as margin over feed costs or margin over pig place per year.

In this context, the principal advantage in *ad libitum* feeding is a significant reduction in fattening time (between two and four weeks) allowing a greater return on investments. If the often narrow range of prices between different commercial classes of carcass is taken into account, economics may well favour a liberal level of feeding (even with fatter types such as castrates) despite the slight reduction in carcass quality that will result.

Mean recommendations Mean recommendations for dietary energy levels as a function of liveweight are given in *Table 11* for those conditions optimal for the production of lean meat (i.e. with the use of females or entire males) from those genotypes usually found in France. They represent a level approximately three times that of maintenance throughout the growing/finishing period.

Table 11 Average recommended energy allowances for gilts and entire males during growing/finishing

Liveweight (kg)	25	30	40	50	60	70	80	90	100–110
DE (kcal/d)	4200	5000	6000	7000	8000	8800	9200	9600	10 000
Food intake (kg/d)*	1.3	1.55	1.9	2.2	2.5	2.75	2.9	3.0	3.1

*Feed based on cereals, with a dry matter content of 87% and a DE value of 3200 kcal/kg. Quantities of food allowed need to be altered with changes in DE content.

Variations to be noted Away from these mean norms, there may be variations based upon the following criteria:

(1) *Type of pig used* As indicated above, those pigs with high muscle development (i.e. Belgian Landrace, Pietrain) may be fed liberally without a reduction in carcass quality.

Castrate males on the other hand require a more severe degree of feed restriction than with females or entire males, particularly during the finishing phase: daily energy intakes may level off at 8000 kcal DE/day (2.5 kg of a feed containing 3200 kcal DE/kg) when the liveweight reaches 60 kg. The leaner carcasses produced are however associated with considerably poorer feed conversion ratios and a significant reduction in rate of growth. In effect, therefore, the use of castrate males for the production of lean meat is hardly compatible with optimal feed utilization and it may be noted that the production of carcasses with only a moderate fat content and at a minimum cost implies the use of animals with high potential for muscle development.

(2) *Constraints imposed by carcass characteristics* The need to provide leaner carcasses requires a more severe degree of feed restriction.

(3) *Slaughter weights* Feed restriction progressively imposed, which limits fat growth with age, is more important the heavier the carcasses.

(4) *Environmental climatic conditions* It may be noted that for a given level of feed intake, daily liveweight gain is reduced by around 16 g/day for each 1°C reduction in ambient temperature in the range 20–10°C. This reduction may be prevented by a supplementary feed level of 35–40 g/day at a dietary energy level of 3200 kcal DE/kg. Taking into account seasonal climatic variations currently experienced in France, the daily feed intake may be increased by approximately 5% in winter.

Protein and amino acids

Mean recommendations

As with energy, mean recommendations concerning dietary levels of protein and amino acids given in *Tables 6* and *12* are designed to cover the requirements of those animals with a high production of lean meat. They are relevant to those animals with the highest requirements (lean female pigs with a rapid growth rate) fed *ad libitum* or with only a moderate level of restriction (a minimum of 90% of *ad libitum* intake) and are not applicable to mixed populations of females and castrate males (which have lower requirements) as is the case with other tables of recommendations (NRC, 1979).

Table 12 Recommended allowances for protein and amino acids (in g/1000 kcal DE)*

	Young pig		Growing/finishing	
	Suckling	Weaner	Growing	Finishing
Liveweight (kg)	5–10	10–25	25–60	60–100
Crude protein:				
Suggested level	63	54	53	47
Minimum level of ideal protein	57	50	45	40
Amino acids:				
Lysine	4.0	3.0	2.5	2.2
Methionine + cystine	2.3	1.8	1.5	1.3
Tryptophan	0.7	0.6	0.45	0.4
Threonine	2.3	1.8	1.5	1.3
Leucine	2.8	2.3	1.8	1.6
Isoleucine	2.3	1.8	1.5	1.3
Valine	2.6	2.0	1.75	1.55
Histidine	1.0	0.8	0.65	0.55
Arginine	1.0	0.9	0.75	0.65
Phenylalanine + tyrosine	3.7	2.8	2.5	2.2

*Multiply these levels by 1/0.95 or 1.05 to convert to g/1000 kcal ME.

Balance between amino acids

The relatively constant composition of protein deposited during growth in the pig means that the necessary quantities of each essential amino acid increase with age at a similar ratio to each other. This moreover is in agreement with the hypothesis on which the establishment of requirements in America (NRC, 1979) and Great Britain (ARC, 1981) is based.

For a lysine requirement equal to 100, the relative requirements of the others may be calculated thus:

Lysine	100
Methionine and cystine	60
Tryptophan	18
Threonine	60
Leucine	72
Isoleucine	60
Valine	70
Histidine	26
Arginine	29
Phenylalanine + tyrosine	100

Recommended dietary levels may vary in terms of absolute values as a function of the importance of lean tissue deposition; however the same balance between all values should always be maintained if possible. Thus values for amino acid levels in France and Europe are together higher than those in the USA (NRC, 1979) because the latter are applicable to pigs fed *ad libitum* with a greater potential for fat deposition. The recommended dietary level for example of lysine in Europe is 2.5 g/1000 kcal DE (liveweight range 25–60 kg) whereas those in the USA are only 2.1 g which is some 20% less. The same relative difference is applicable to all amino acids to maintain the balance between them.

Among those semi-essential amino acids, cystine and tyrosine will more or less meet half the requirements for sulphur (methionine + cystine) and aromatic (phenylalanine + tyrosine) amino acids, respectively. In the nutrition of the pig, only the requirement for total sulphur amino acids is taken into consideration.

Dietary protein levels

There are two recommended dietary protein levels (*Tables 6 and 12*):

(1) *The suggested level* for the most commonly used diets (based on cereals and soya-bean meal) correctly balanced for amino acids. In order to meet the required level of lysine, using protein sources in which it is deficient, there is an oversupply in the amount of others.
(2) *Minimum level of balanced protein* This is related to the minimum level of protein sources required to cover protein needs, which is the same as that when requirements for all essential amino acids have been met. Under practical conditions (diets based on cereals) this is the level obtained following free lysine supplementation (first limiting factor). The level of lysine in balanced protein is of the order of 5.5%.

The difference between these two levels is associated with the possibility of reducing protein levels with the use of synthetic lysine: with those diets currently used based on cereals, this reduction is of the order of 2%, but it is all the more important when the protein content of the cereal in question increases as this is associated with a relative reduction in its lysine content (e.g. wheat).

Balance between protein (or amino acids) and energy

Dietary protein and amino acid levels must be adjusted as a function of energy content as indicated in *Table 12*.

For example, between 25 and 60 kg liveweight, the recommended level of lysine would be:

2.5 × 3.2 = 8.0 g/kg for an energy level of 3200 kcal DE/kg
2.5 × 3.4 = 8.5 g/kg for an energy level of 3400 kcal DE/kg

Factors influencing protein requirements

It is apparent from the above discussion that an increase in the potential for muscle development is associated with opposing changes in optimum daily amounts of energy and protein/amino acids: requirements for the former fall and those for the latter increase in this situation. This raises the margin of variation in the relative requirements of protein/amino acids expressed as a ratio to energy.

Effect of sex and genotype

Differences in amino acid requirements, according to the ranges in muscle development found between genotypes, have been insufficiently quantified in pigs to allow them to be taken into consideration.

When considering sex, it is appreciated that protein requirements at the same energy level for females are around 10% higher than in castrate males, principally during the finishing phase (60–100 kg liveweight); this is particularly so for lysine as for total protein requirements (*Table 13*).

Table 13 Influence of sex on relative protein and lysine requirements during growing/finishing

	Liveweight (kg)	
	25–60	60–100
Lysine requirements (g/1000 kcal DE):		
Gilts	2.5	2.2
Castrates	2.5	2.0
Total ideal protein requirements (g/1000 kcal DE):		
Gilts	45(39)*	40(35)
Castrates	45(39)*	36(31)

*Being expressed as, respectively, total crude protein and, in brackets, digestible crude protein (assuming an apparent digestibility of 87% for a diet on maize/soya-bean meal).

However, when castrate males are restricted fed during finishing (75–80% of *ad libitum*) in order to limit fattening, the relative requirements for protein/amino acids are higher in comparison with those calculated for *ad libitum* feeding: mean values thus proposed are particularly relevant to this case.

It is generally accepted that amino acid requirements appropriate to females may be equally applied to entire males. It is possible however that the requirements of entire males are slightly higher, although this has not been established definitively. Therefore, to provide a margin of security, lysine levels of 0.9–1.0% and 0.8% respectively for growth and finishing in diets containing 3200 kcal DE/kg (giving 2.7 and 2.5 g lysine/1000 kcal DE) could be supplied. The percentages of other essential amino acids would increase accordingly to maintain the balance with lysine depleted earlier, and the suggested level of protein would be 18% and 16% respectively for growth and finishing.

Effect of level of feeding, or energy level

The recommended dietary energy level for suckler diets is optimal (*Table 6*). The level of protein may vary independently of this according to whether there is an ideal balance between all the essential amino acids (as a percentage of protein or as a ratio to lysine). For post-weaning diets, a variation in the dietary energy level is accompanied by alterations in the levels of essential amino acids associated with their ratio to the concentration of DE or ME.

There is only limited information available on the variation in protein requirements as a function of the level of energy restriction in growing/finishing pigs. It may be suggested that a moderate level of restriction (5–10% in the case of females) is not associated with a significant alteration in daily amino acid requirements; accordingly, no difference in relative protein levels (as a percentage of feed or as a ratio to energy) between feeding *ad libitum* or with a small degree of restriction is suggested. The margins of security included in the recommendations are sufficient.

However, this does not follow for more severe levels of restriction (20–25% as in the case of castrate males); here daily amino acid requirements are reduced but not to the same degree as the level of restriction with the result that there is a relative increase in requirements to a level that is unfortunately not yet precisely known. In the absence of accurate information relating to this subject it is recommended that, whatever the conditions, a daily amount of protein and amino acids as given in *Table 14* be employed. This is particularly so for lysine, which is the amino acid most frequently limiting.

Table 14 Recommended daily allowances of lysine and crude protein for finishing pigs

	Liveweight (kg)					
	20	*40*	*60*	*80*	*100*	*120*
Lysine (g/d)	10	14	17	20	21	22
Crude protein (g/d):						
Usual quantity	180	250	300	360	380	400
Ideal protein*	155	220	260	315	330	350

*Allowances of essential amino acids (the first being lysine) are covered. Quantities are estimated from a diet based on maize/soya-bean meal with an assumed apparent digestibility for crude protein of 87%.

Dietary fibre level and amino acid availability

The recommendations given for growing/finishing pigs are appropriate for a feed based on a mixture of cereals (maize, wheat, barley) and soya-bean meal with a protein digestibility of 86–87%. When considering diets of lower dietary energy value higher in fibre a simple calculation must be made as a consequence of these differing values: thus it is known that the presence of fibre reduces the availability of protein and amino acids although by an as yet unknown amount. In order to avoid a risk of deficiency of certain amino acids a compensatory increase in dietary levels of protein and amino acids relative to energy may be suggested. As a guideline, a fall of 1–1.5% in apparent protein digestibility with each 1% increase in dietary crude fibre is possible (on a dry matter basis). Of course this correction can only be considered an approximation. There is as yet no means of expressing dietary levels of protein and amino acids for the pig in terms of their availability.

Use of only one diet during growing/finishing

With the use of only one diet throughout the fattening period (25–100 kg liveweight) those recommendations established for the liveweight range 20–60 kg must be adopted, although dietary levels of protein and amino acids will be in excess of requirements during the finishing phase. Energy provided by excess protein is however utilized for growth although with a lower efficiency; this may reduce the level of fat in these animals. On the other hand it is known that the reduction in protein requirements during finishing (relative to energy) is less apparent with certain genotypes with a high potential for muscle development. As a result, wastage of protein as a consequence of using only one diet during the entire period of fattening is more a problem with fat or castrate male pigs, and is less so with lean pigs and females.

Minerals and vitamins

Phosphorus, calcium and sodium are those elements most frequently lacking in pig diets. Magnesium and sulphur, which are sometimes deficient in ruminant diets, are usually adequate, while chloride and potassium are rarely sub-optimal.

Calcium and phosphorus

As indicated in Chapter 4, recommended dietary levels of calcium and phosphorus are based on factorial estimates of their requirements. It may be recalled that this method is based upon assessments of net requirements for each physiological process (maintenance, growth, gestation, lactation) and by subsequently multiplying them by a factor for the coefficient of digestibility (100/CTD) which accounts for true intestinal absorption.

This method must be accompanied by a careful choice of criteria used to estimate net calcium and phosphorus requirements. Thus many easily measurable criteria (liveweight gain, feed conversion ratio and plasma phosphorus and calcium together with phosphatase activity) are in pigs too sensitive to variations in the balance of phosphorus to calcium. On the other hand, parameters including bone density and its resistance to breakage, satisfactorily reflect the state of mineralization of the skeleton.

The theoretical bases adopted for the calculation of phosphorus and calcium requirements for the growing pigs are presented in *Table 15*.

(1) *Net maintenance requirements* correspond to involuntary losses in the faeces and urine. Minimum values have been determined from studies with radio-isotopes to be 35 mg of calcium/kg liveweight per day, of which only 3 mg is lost via the urine, and an average figure of 20 mg for phosphorus of which half is excreted in the urine.

(2) *Net requirements for growth* correspond to fixed quantities of calcium and phosphorus which assure optimal mineralization and liveweight gain. These amounts are more difficult to define because they are dependent both upon the required rate of gain and degree of mineralization. Thus for pigs slaughtered at 100 kg liveweight, a heavily mineralized skeleton is not essential as long as there is sufficient resistance to bone breakage. For modern types of pig, 7–10 g of calcium and 5–7 g of phosphorus per kg of gain are necessary to promote a sufficient level of mineralization.

The calculation of dietary requirements by the factorial method is based ultimately on mean values for the coefficient of true digestibility (CTD) for mineral elements. Such an approach is now possible for calcium and phosphorus as a consequence of much work carried out in pigs with the help of ^{32}P and ^{45}Ca,

Minerals and vitamins

Table 15 Basis for the calculation of calcium and phosphorus allowances for the growing pig

	Calcium	Phosphorus	Comments
Daily net maintenance requirements per kg liveweight	35 mg	20 mg	Assuming a urinary loss of 3 mg Ca and 10 mg P
Net requirements for growth per kg liveweight gain	11 g 10 g 9–9.5 g 8 g 7 g 5 g	7 g 6 g 5.5–6 g 5.5 g 5.5 g 3 g	Liveweight: Up to 20 kg 20–50 kg 50–100 kg 100 kg Above 100 kg For maternal weight gain of gestating sows
True digestibility (%)	70 60 55 50 45	60 55 50 50 50	Liveweight: 5 kg 10 kg 20 kg 35 kg 50 kg and above

although it is difficult for other major elements (magnesium, sodium and potassium) and practically impossible for trace elements.

The CTD for calcium varies between 45 and 55% except for the very young piglet where it may pass 60%. Like other species, the pig is in addition able to modify the efficiency of intestinal absorption of calcium based upon the level in the diet. If this is very low (intake of 2 g/day) the CTD may reach 80%. However, under practical feeding conditions, this adaptive mechanism should not be relied upon since minimum daily net calcium requirements are between 6 and 7 g beyond 20 kg liveweight.

The CTD for phosphorus varies considerably as a function of its dietary form. Thus phosphorus from good-quality mineral phosphates has an absorption higher than 65% whereas only 30–35% is absorbed from phytate phosphorus which is abundant in cereals and oil meals. Taking a dietary level of phosphorus of 6 g/kg, of which 4 g is supplied by cereals and oil meals and 2 g by mineral phosphates, it may be estimated that half the phosphorus is present in phytate form and the other half in mineral form. For a respective CTD of 35 and 70% the mean CTD of this typical diet would thus be in the region of 50% in pigs above 20 kg (a value confirmed by many investigations) and slightly higher in piglets (55–60%).

For pigs, the dietary levels of phosphorus are, as has already been mentioned, expressed as total (that revealed by analysis) rather than available as it would seem premature under practical feeding conditions to take into account variations in availability of phosphorus from plant sources.

Recommended dietary levels of phosphorus presented have been established on the basis of the utilization of mineral phosphates of suitable quality (with a minimum level of phosphorus extracted with citric acid of 85%) and only with the use of very poor-quality sources would the values given need to be increased.

Recommended dietary levels of calcium and phosphorus calculated according to the above considerations are detailed in *Table 16* for pigs at different stages of growth. They are expressed in terms of daily amounts (g) and as a dietary concentration (g/kg).

These values take into account current practical conditions found during the rearing of pigs (liveweight, average liveweight gain, level of feeding). Higher levels of both calcium and phosphorus would be necessary to promote maximal bone mineralization but this would not be accompanied by any improvement in performance or by an increased bone strength. Finally, levels of calcium and phosphorus higher than those recommended would not reduce the incidence of osteochondrosis or of hoof weakness.

Table 16 Recommended allowances for calcium and phosphorus in the young and growing/finishing pig

Liveweight (kg)	<10	10	20	35	50	70	100
Daily liveweight gain (g)*	250	350	500	600	750	800	900
Daily feed intake (g)**	0.35	0.6	1.1	1.6	2.1	2.5	2.8
Calcium (g/d)	4.5	7.0	10.5	15.0	20	21	24
Phosphorus (g/d)	3.2	5.0	8.0	9.5	11	12	14
Calcium (g/kg feed)**	13.0	11.5	9.5	9.5	9.5	8.5	8.5
Phosphorus (g/kg feed)**	9.0	8.0	7.0	6.0	5.0	5.0	5.0

*Corresponding to an average daily liveweight gain of 750 g between 20 and 100 kg live weight.
**Air-dry feed.

These recommendations must nevertheless be modified as a function of:

(1) *Level of feed used* For this it is sufficient to take into account the daily requirements given in tables mentioned above.
(2) *Performance achieved* As requirements for phosphorus and calcium are a function of the rate of growth of animals, mineral levels for growth must be increased for pigs reared under very intensive conditions (10–12 g of calcium and 7 g of phosphorus/kg feed) although for traditional rearing with lower output 8 g of calcium and 5 g of phosphorus/kg feed should be sufficient.

Finally, as the number of feeds is in practice necessarily limited, it may not be possible accurately to adjust the dietary levels according to the requirements of the animal at each stage of growth. Therefore it is necessary to choose mean levels for each category as detailed in *Table 6*, occasionally with a slight increase over the actual requirements.

Magnesium, sodium and trace elements

The minimum magnesium requirement is of the order of 0.4 g/kg feed. It is invariably met in typical diets which have levels higher than 1 g/kg (dry matter basis). It is therefore not advised to supplement diets with magnesium. Sodium requirements are estimated to be 1.5 g/kg feed on a dry matter basis and are met by a dietary sodium chloride level of 0.5%.

For the many reasons discussed in Chapter 4, recommended dietary levels of trace elements are associated inevitably with a degree of imprecision. Methods of determination of requirements are in effect often very general and the variation in levels in raw materials difficult to estimate. Furthermore there are problems associated with availability and the numerous interactions between minerals.

Risks of deficiency of trace elements exist in the pig for manganese, copper, zinc, iodine and selenium. Iron is only lacking in diets for young piglets. Among essential trace elements, selenium, fluorine and molybdenum may equally pose problems of toxicity.

Recommended levels, expressed in terms of mg/kg feed (or ppm) are presented in *Table 7*. They are generally much higher than those needed to meet requirements and take into account the most unfavourable conditions of feed utilization.

Vitamins

Recommended added amounts of vitamins in feeds for the pig are presented in *Table 7*. They take into account vitamin levels in those raw materials generally employed and include a margin of security sufficient for commonly used diets. The use of novel raw materials or of new diet formulations must however be accompanied by an assessment of the adequacy of total vitamin levels.

Water

Water is the most important element of the daily diet. It is essential even for starved animals as it allows the excretion of metabolic waste-products. With an adequately balanced diet, and within the thermoneutral zone, water consumption by the weaner piglet is between 3 and 3.5 litres/kg dry matter consumed which will account for all requirements. This is reduced to between 2.2 and 2.5 litres/kg for growing/finishing pigs. Under no circumstances, whatever the category of pig in question, is this less than 2 litres/kg. With growing pigs, an intake higher than 4 litres/kg is excessive and may have a depressive effect upon growth rate. Requirements are not higher than this except in diets with excessive levels of minerals, particularly potassium and sodium (if molasses or whey are used), when they may reach 5 or 6 litres/kg with considerable individual variations. Requirements may equally increase with a sudden and significant increase in temperature before long-term adaptive mechanisms are brought into operation. Amounts may reach 4–5 litres/kg in the growing pig.

Taking into account these variations due to feeding, environment and individual variability it is recommended that water be provided *ad libitum*. If this is not practised, then a minimum of 1.5–2 litres/day for the weaner piglet and 4–7 litres for fattening pigs should be allowed.

Chapter 7
Nutrition of breeding pigs

The breeding herd may be divided into four categories:
(1) Gilts up until entry into the herd (110–130 kg liveweight).
(2) Young and adult boars.
(3) Sows — initial or subsequent pregnancy.
(4) Sows — initial or subsequent lactation.

Nutrition of gilts and boars

Gilts

The usual energy levels for gilts are comparable with those established for fattening pigs. As with the latter, the level of feeding adopted must allow an optimal development of lean tissue while limiting the amount of fat deposited; this may be achieved by a progressive degree of feed restriction up to a daily intake of between 9000 and 9500 kcal DE until 100 kg liveweight or around 180 days of age, corresponding to approximately 3 kg of feed of 3000 kcal/kg DE. The level of feeding for a given liveweight must be between 85 and 100% of *ad libitum* intake.

With the approach of puberty (which is at 200–220 days of age at a liveweight of 100–130 kg) gilts must be restricted more severely. For this, the level applied at 100 kg liveweight is progressively reduced to between 7500 and 8000 kcal DE/day. Only a particularly severe degree of feed restriction (less than 70% of *ad libitum*) is capable of significantly delaying the onset of puberty. However the resultant reduction in growth rate is also associated with lower ovulation rates, levels of fertility at the first heat and of milk production during the subsequent lactation. Similarly these responses are related to a restriction in protein intake or a deficiency in a limiting amino acid.

The almost complete absence of experimental data on the effect of nutrition of gilts at the end of rearing on their subsequent reproductive performance means that it is not currently possible to give precise recommendations concerning their energy requirements. Unfortunately the same is true for dietary levels of protein and amino acids, and in the absence of available information those recommendations applicable to growing/finishing pigs are proposed. Recommended dietary mineral levels are 9 g for calcium and 6 g for phosphorus/kg feed for a gilt consuming around 2.5 kg daily.

Young and adult boars

There is no information whatsoever concerning the dietary requirements of boars both young and old.

Consequently, as with gilts, the recommendations that are used with growing/finishing pigs in terms of energy, protein and amino acids are applied equally to young boars. Taking into account those conclusions outlined above, relevant to the entire male (*see* page 40), a slight increase in the levels of protein and amino acids relative to energy may be recommended for young boars. Additionally, and outside of those effects recorded for growth, a restriction in energy or protein levels will only adversely influence fertility in terms of attainment of puberty and sperm quality if it is severe.

With adult boars (from two years of age) optimal energy levels must cover, essentially, maintenance requirements together with those associated with physical exercise, climatic conditions (rearing outside or semi-extensively) or high frequency of service. The level used may be in the region of 7500–8000 kcal DE/day (the same as with the pregnant sow), which corresponds to an intake of 2.5–2.7 kg of feed with an energy level of 3000 kcal DE/kg.

Recommended levels of protein and amino acids may be arrived at in the same way as with the pregnant sow (considered subsequently in this chapter, page 59). An increase in the frequency of usage (either natural service or semen collection) may be associated with a slight increase in sulphur amino acid requirements (due to the relatively high concentration of cystine in sperm) although this remains to be verified. Dietary levels of 7 g of calcium and 5 g of phosphorus/kg feed are adequate for the boar. Its relatively low water consumption (2.0–2.2 litres/kg dry matter consumed) is met by the provision of 7 litres/day.

Nutrition of sows

Average recommended levels of energy, protein, amino acids and macroelements are given in *Table 17*. With the exception of energy, they are expressed in terms of a percentage of the diet of known energy level. Recommended levels of trace elements together with suggested supplementary amounts of vitamins are presented in *Table 18*.

Energy

Energy expenditure of sows varies according to the reproductive cycle (pregnancy, lactation, post-weaning).

During pregnancy

Energy requirements of the pregnant sow are associated primarily with maintenance (65–70% of total requirements in contrast to 30–35% with the growing pig) together with those related to the formation of:

(1) Maternal tissue (i.e. mammary tissue).
(2) The conceptus (fetus, associated embryonic tissues and liquids), the development of which is particularly important during the last third of the pregnancy

Table 17 Recommended allowances for energy, protein, amino acids and minerals in breeding sows

	Gestation*	Lactation
Dietary energy level (kcal DE/kg feed)**:		
Range†	2800–3300	3000–3300
Average†	3000	3100
Crude protein (% feed):		
Usual quantity	12	14
Amino acids (% feed):		
Lysine	0.40	0.60
Methionine + cystine	0.27	0.33
Tryptophan	0.07	0.12
Threonine	0.34	0.42
Leucine	0.30	0.69
Isoleucine	0.34	0.42
Valine	0.43	0.42
Histidine	0.12	0.23
Arginine	–	0.40
Phenylalanine + tyrosine	0.31	0.69
Minerals (% feed):		
Calcium	1.00	0.80
Phosphorus	0.55	0.55
Daily feed intake (kg/d)	2.5	4.5–5.5
Daily energy allowance (kcal DE/d)	7500	14 000–17 000

*Appropriate for boars.
**Dry matter of diets is 87%.
†Or, using ME (where ME=0.95 DE) a mean concentration of 2850 kcal/kg for gestating sows (range 2660–3130) and 2940 kcal/kg for lactating sows (range 2850–3130).

and which corresponds to 450–500 kcal deposited daily after 100 days of pregnancy, in contrast to 120 kcal at 70 days and 60 kcal at 40 days.

The optimal recommended dietary energy level during pregnancy varies according to the criterion under consideration. Thus a significant nutrient reduction has no effect on the pregnancy itself but may considerably reduce the deposition of maternal tissue, the subsequent synthesis of the constituents of milk and therefore the growth rate of the piglets. Energy levels during pregnancy must therefore be fixed at that level sufficient to promote a net maternal gain during gestation (the gain following weight loss associated with farrowing) compatible with optimal subsequent performance. This level of gain may be fixed at 30 kg and is satisfactorily achieved with a daily energy intake of around 7500 kcal DE throughout pregnancy. There is no information available which allows the calculation of energy requirements according to parity. Consequently, as a first approximation the same recommendations for multiparous as for primiparous sows are used, as the additional requirements associated with growth in the latter are compensated for by lower maintenance needs.

In the same way, variations in energy requirements, as a consequence of physical activity (means of keeping animals) or thermal environment (climate, season) are poorly understood. As with growing pigs an increase of around 5% over usual daily requirements may be recommended during winter.

Table 18 Recommended added dietary levels of trace elements and vitamins in the sow (in iu/kg or mg/kg)

Trace elements:			
Iron	80	Cobalt	0.1
Copper	10	Selenium	0.1
Zinc	100	Iodine	0.6
Manganese	40		
Fat-soluble vitamins:			
Vitamin A (iu)	5000	Vitamin E (mg)	10
Vitamin D (iu)	1000	Vitamin K_3 (mg)	0.5
Water-soluble vitamins:			
Thiamine	1	Biotin	0.1
Riboflavin	3	Folic acid	0.5
Calcium pantothenate	8	B_{12}	0.02
Niacin	10	Choline chloride	500

A wide variability in dietary energy levels (between 2800 and 3200 kcal DE/kg) is possible. The relatively low energy requirement of the sow compared with its potential feed intake capacity allows, additionally, the use of feeds comparatively low in energy and this is the more so as the sow is better able to utilize fibrous diets than the young pig. Furthermore, a significant intake of these low-energy feeds makes the sow more content. Finally, bulky diets stimulate intestinal peristalsis and help to reduce risks of constipation towards the end of the pregnancy. For a feed with an energy level of 3000 kcal DE/kg, the mean daily intake is of the order of 2.5 kg. As indicated above, intake of each animal must be adjusted to achieve the net required gain.

It is difficult to know whether it is worthwhile increasing feed intake during the last third of pregnancy (after day 70) to accompany the rapid growth of the conceptus. It has been shown that a given net gain is achieved more by the total amount of food consumed during pregnancy than by variations in intake during this period. However it is possible that, following the results of recent studies, an increase in feeding level towards the end of pregnancy may improve piglet post-natal survival.

During lactation

Other than for maintenance, energy costs during lactation are associated with the synthesis of the constituents of milk. For a mean daily milk production of 6–7 kg of milk, the amount of energy lost is 7000–8000 kcal (1050–1150 kcal/kg). It is known that milk production increases with litter size but only slightly; the amount of milk available for each piglet is therefore less with larger litters.

As a consequence of the high level of energy costs during lactation, the high-yielding sow is unable to consume sufficient food to meet these costs even under *ad libitum* feeding conditions. As a result there is weight loss during lactation, which may be of the order of 10–25 kg, associated not only with the mobilization of extra-uterine tissue but with water loss and involution of the uterus.

This weight loss and therefore tissue mobilization varies as a function of the prolificacy of the sow (pregnancy number and its size) and nutritional state during pregnancy: it is greater the higher the net gain (of body reserves) during pregnancy. Conversely, it is evidently lower the earlier the piglets are weaned.

Following from this inevitable mobilization of body reserves (principally at the beginning of lactation when appetite of sows is reduced), it is recommended that lactating sows with ten or more piglets should be fed *ad libitum* a standard diet with a mean energy level of between 3000 and 3200 kcal DE/kg. The average daily feed intake will be in the region of 4.5–5.5 kg corresponding to an intake of 14 000–17 000 kcal DE on a diet of 3100 kcal DE/kg.

For smaller litters (less than eight piglets) this level of feed is corrected by subtracting 850 kcal DE (or 275 g of feed of 3100 kcal DE/kg) per piglet.

Protein and amino acids

During pregnancy

There have only been a few studies to evaluate amino acid requirements of pregnant sows, and these have not been definitive because of:

(1) too short a time period employed;
(2) the use of criteria (protein levels at the end of pregnancy, plasma-free amino acids), which do not directly measure reproductive performance;
(3) too few animals used;
(4) the use of reduced litter size (usually seven to eight piglets);
(5) the absence of measurements of long-term effects on overall sow productivity (body reserves, longevity) and piglet growth rate.

As a result, the available recommendations (NRC, 1979; ARC, 1981) can only be considered as minimal. Actual protein and essential amino acid requirements of very prolific sows of all ages have yet to be established and need to include both the current reproductive state of the animal and its future long-term performance.

Daily dietary protein and amino acid levels must allow for nitrogen deposition in the uterus and associated tissue (slightly less than 3 g/day) together with maternal tissue (between 6 and 12 g deposited daily). In multiparous sows maternal deposition corresponds to a liveweight increase of around 30 kg, of which a significant proportion is adipose tissue. In primiparous sows, entering the breeding herd at a relatively early age, dietary nitrogen levels must be adjusted to take into account a net gain in pregnancy of between 30 and 40 kg, of which a proportion is lean tissue associated with growth.

Taking into account these factors, together with the increase in sow productivity during the last few years, the recommended dietary levels of amino acids for pregnant sows (*Table 19*) have been systematically increased over those used previously by the NRC (1979) and ARC (1981). Therefore a daily lysine intake of 10 g has been proposed which corresponds to a daily intake of 2.5 kg of a feed, containing 3000 kcal DE/kg, with a lysine level of 0.4%. Percentages of other amino acids have been calculated on the basis of their relative values compared to lysine proposed by the ARC (similar to those adopted by the NRC).

The overall protein level, for the amount of feed considered above, is 12% which corresponds to a daily intake of around 250 g (compared with 216 g and 180 g from

Table 19 Recommended daily allowances for amino acids for the gestating sow (g)

Amino acids	NRC (1979)		ARC (1981)		Recommendation (g/d)
	g/d	Balanced protein (lysine=100)	g/d	Balanced protein (lysine=100)	
Lysine	7.7	100	8.6	100	10
Methionine + cystine*	4.1	53	5.8	67	6.7
Tryptophan	1.6	21	1.4	16	1.6
Threonine	6.1	79	7.2	84	8.4
Leucine	7.6	98	6.4	74	7.4
Isoleucine	6.7	86	7.4	86	8.6
Valine	8.3	107	9.2	107	10.7
Histidine	2.7	35	2.6	30	3.0
Arginine**	0	–	0	–	0
Phenylalanine + tyrosine*	9.4	121	6.6	77	7.7

*As with the growing pig, provisions of cystine and tyrosine enable about half the allowances for, respectively, methionine + cystine and phenylalanine + tyrosine to be met. This is equally valid for the lactating sow.
**Arginine does not appear to be an essential amino acid in adult pigs or pregnant sows.

the NRC and ARC respectively). The possibility of supplementation of diets with low protein levels with free amino acids is uncertain and a minimum level of balanced protein is not proposed.

During lactation

Protein and amino acid requirements during lactation are based above all upon the level of milk produced (itself based upon the litter number and size) and the length of lactation: a sow producing 6–7 kg of milk daily, with a protein content of 5.8% containing 7.75% lysine, therefore loses 350–400 g of protein and 27–32 g of lysine daily. However, protein levels during lactation must equally take into account nutritional state during pregnancy since this influences the increase in body reserves which are subsequently available for the synthesis of the constituents of milk.

Experiments conducted to determine the amino acid requirements of lactating sows suffer from the same faults as those with pregnant animals. It follows in the same way that estimations of requirements determined up until now risk being marginal under certain rearing conditions.

However, in the absence of other information, those daily recommendations during lactation proposed by the ARC (1981) have been adopted (*Table 20*) and are for lysine, the most limiting amino acid, 33 g. With sows at the peak of their production, consuming 5.5–6.0 kg daily of a feed of 3100 kcal DE/kg a dietary level of 0.6% lysine is therefore necessary. The figures recommended presented in *Tables 17* to *20* have been based upon the balance of amino acids to lysine proposed by the ARC (1981). It may be noted that this balance corresponds very closely to the amino acid profile of milk protein.

Table 20 Recommended daily amino acid allowances for the lactating sow (g)

Amino acids	NRC (1979)*		ARC (1981)	
	g/d	Balanced protein (lysine=100)	g/d	Balanced protein (lysine=100)
Lysine	31.9	100	33	100
Methionine + cystine	19.8	62	18	55
Tryptophan	6.6	21	6.3	19
Threonine	23.6	74	23	70
Leucine	38.5	121	38	115
Isoleucine	21.4	67	23	70
Valine	30.2	95	23	70
Histidine	13.8	43	13	39
Arginine	22.0	69	22	67
Phenylalanine + tyrosine	46.8	147	38	115

*Based on a daily DE intake of 18 700 kcal.

Estimates of total protein requirements, which may be determined factorially or by direct measurements, suggest recommended daily intakes of 700 and 800 g for primiparous (producing 5.5 kg of milk) and multiparous sows (producing 7 kg of milk) respectively — both of which may be achieved with a dietary protein level of 14%.

As indicated with the pregnant sow, the possibilities offered by the use of free amino acids are too uncertain to allow the recommendation of a minimum level of balanced protein.

Above all, dietary levels during lactation, irrespective of whether energy or amino acids are considered, cannot be dissociated from those that operate during pregnancy. Nutrient requirements throughout the entire reproductive cycle depend initially upon the amount of milk secreted and the length of lactation. However, alterations in dietary levels during pregnancy help to meet these demands. The possibility of such changes is justification for the use of only one diet throughout the reproductive cycle with a protein level between 13 and 14% (instead of 12% only during pregnancy). This technique is only considered if excessive mobilization of body reserves during lactation, during a short lactation of 2–3 weeks or with an average litter size of 8–9 piglets, is to be avoided.

Minerals, vitamins and water

The bases of the calculation of calcium and phosphorus requirements of pigs using the factorial method are considered in the previous chapter. Those for the sow, in quantities per day, are indicated in *Table 21*.

The following may be noted:

(1) There is a requirement of 5 g of calcium and 3 g of phosphorus per kg of maternal liveweight gain excluding the conceptus.
(2) Net requirements for pregnancy are negligible during the initial two months as bone mineralization of the fetuses occurs principally during the last third. The evaluation of net requirements is based on the mineral composition of the

new-born piglet (10 g of calcium and 6 g of phosphorus per kg) and litter weight (being 12 kg and 14 kg respectively for primiparous and multiparous sows).
(3) Net lactation requirements are associated with a composition in the milk of 2.0 g and 1.5 g of calcium and phosphorus respectively for a lactation of 4–5 weeks.

Table 21 Basis for the calculation of calcium and phosphorus requirements of sows

	Calcium	Phosphorus	Comments
Daily net maintenance requirements per kg liveweight	35 mg	20 mg	Assuming a urinary loss of 3 mg Ca and 10 mg P
Daily net requirements during the final third of pregnancy	3 g 3.7 g	2 g 2.2 g	First pregnancy Subsequent pregnancies
Net lactation requirements per kg milk	2 g	1.5 g	Lactation of 3–4 weeks
True digestibility (%)	45 55	50 50	Gestation Lactation

Taking account of these figures, and the average values for CTD (*Table 21*), it is possible to calculate recommended dietary levels of calcium and phosphorus which are given in *Table 22*. Levels of trace elements and vitamins are given in *Table 18*.

Table 22 Recommended allowances for calcium and phosphorus in sows

	Maintenance and for the initial two-thirds of pregnancy*	Final third of pregnancy*	Lactation*
Liveweight (kg)	130–180	160–200	140–180
Daily liveweight gain (g)**	300–400	200–250	–
Daily milk production (kg)	–	–	67
Daily feed intake (kg)†	2.5	2.5	4.5–5.0
Calcium (g/d)	16.0–17.5	23–26	34–41
Phosphorus (g/d feed)	9.0	12–14	22–27
Calcium (g/kg feed)†	6.5–7.0	9.5–10.5	7.5–8.0
Phosphorus (g/kg feed)†	4.0	5.0–5.5	5.0–5.5

*First values refer to the initial gestation/lactation; second values refer to subsequent gestations/lactations.
**Maternal liveweight (excluding that of the conceptus).
†Air-dry feed.

Among pigs, sows have the highest water requirements and consumption is between 4.0 and 4.5 litres/kg dry matter consumed, with the highest levels required by the lactating sow. If water is not provided *ad libitum* a minimum of 10 litres/day and 20 litres/day respectively for pregnant and lactating sows has to be provided and these amounts must be increased in times of excessive heat or with the use of poorly balanced by-products.

Chapter 8
Nutrition of rabbits

Experimental work conducted in France and throughout the world during the last 15 years or so has allowed the definition of reliable recommendations to be used in the formulation of feeds for rabbit production under temperate European conditions. These recommendations are applied to four categories of rabbit:

(1) adults at maintenance (males, non-pregnant does, those to be culled);
(2) pregnant does (but not lactating);
(3) lactating does (pregnant or not);
(4) young between weaning (around one month) and slaughter at about two and a half months.

The latter are those that have benefited most from the greatest amount of research work.

The recommendations provided have been established under those ambient conditions currently found in Europe and, equally, are based upon the relative costs of feeds in these countries. Under certain local conditions, figures can alter slightly from these norms, thus allowing, possibly, more satisfactory economic results. Limits beyond which it would be unwise to extend, however, will be suggested.

Optimal composition of feeds

When comparing the requirements of the four rabbit categories defined above, it is apparent that lactating does have to receive the most concentrated feeds. In fact, they produce daily between 100 and 300 g of milk, three times more concentrated than that of the cow, and only marginally utilize body reserves in order to achieve this.

Relatively and progressively less concentrated feeds are required by young growing rabbits, does that are pregnant and, finally, males together with non-productive does who may be fed diets of low nutrient density.

Recommended dietary characteristics for each category of rabbit are presented in *Table 23*. In addition to energy, protein, minerals and vitamins, the amount of indigestible crude fibre necessary to provide a minimum amount of gut fill, essential for adequate intestinal function in the rabbit, is provided.

In addition, the table includes the necessary composition of a feed which may be used throughout for all animals: it is a compromise between that necessary for the young growing rabbit and lactating doe. Other categories may be fed this rather

Table 23 Recommended nutrient levels in diets for various categories of rabbits reared intensively

Dietary composition (assuming a dry matter content of 89%)	Units	Growing rabbits (4–12 weeks)	Lactating does and suckling young	Pregnant does but not lactating	Adults at maintenance (bucks etc.)	Complete feed for reproduction + fattening
Digestible energy	kcal/kg	2500	2600	2500	2200	2500
Metabolizable energy	kcal/kg	2400	2500	2400	2120	2410
Fat	%	3	3	3	3	3
Crude fibre	%	14	12	14	15–16	14
Indigestible crude fibre	%	12	10	12	13	12
Crude protein	%	16.0	18.0	16.0	13.0	17.0
Amino acids:						
Lysine	%	0.65	0.75	–	–	0.70
Sulphur amino acids	%	0.60	0.60	–	–	0.60
Tryptophan	%	0.18	0.22	–	–	0.20
Threonine	%	0.55	0.70	–	–	0.60
Leucine	%	1.05	1.25	–	–	1.20
Isoleucine	%	0.60	0.70	–	–	0.65
Valine	%	0.70	0.85	–	–	0.80
Histidine	%	0.35	0.43	–	–	0.40
Arginine	%	0.90	0.80	–	–	0.90
Phenylalanine + tyrosine	%	1.20	1.40	–	–	1.25
Minerals:						
Calcium	%	0.40	1.10	0.80	0.40	1.10
Phosphorus	%	0.30	0.80	0.50	0.30	0.80
Sodium	%	0.30	0.30	0.30	–	0.30
Potassium	%	0.60	0.90	0.90	–	0.90
Chloride	%	0.30	0.30	0.30	–	0.30
Magnesium	%	0.03	0.04	0.04	–	0.04
Sulphur	%	0.04	–	–	–	0.04
Trace elements:						
Iron	mg/kg	50	100	50	50	100
Copper	mg/kg	5	5	–	–	5
Zinc	mg/kg	50	70	70	–	70
Manganese	mg/kg	8.5	2.5	2.5	2.5	8.5
Cobalt	mg/kg	0.1	0.1	–	–	0.1
Iodine	mg/kg	0.2	0.2	0.2	0.2	0.2
Fluorine	mg/kg	0.5	–	–	–	0.5
Vitamins:						
Vitamin A	iu/kg	6000	12 000	12 000	6000	10 000
Vitamin D	iu/kg	900	900	900	900	900
Vitamin E	mg/kg	50	50	50	50	50
Vitamin K	mg/kg	0	2	2	0	2
Vitamin B_1 (thiamine)	mg/kg	2	–	0	0	2
Vitamin B_2 (riboflavin)	mg/kg	6	–	0	0	4
Panthothenic acid	mg/kg	20	–	0	0	20
Vitamin B_6 (pyridoxine)	mg/kg	2	–	0	0	2
Vitamin B_{12}	mg/kg	0.01	0	0	0	0.01
Niacin	mg/kg	50	–	–	–	50
Folic acid	mg/kg	5	–	0	0	5
Biotin	mg/kg	0.2	–	–	–	0.2

concentrated feed without any serious disadvantages. The situations under which either this mixed feed or those more specialized may be employed are considered on page 69. Prior to this it is important to consider nutrient requirements.

Dietary energy level

Both young growing rabbits and female breeding does are able to adjust their feed intake as a function of its dietary energy concentration if the levels of protein and other nutrients are balanced. With the young New Zealand or Californian type, daily energy intake is maintained at around 220–240 kcal DE/kg $W^{0.75}$. In lactating does, it is of the order of 300 kcal DE and may be higher than 360 kcal DE at peak lactation (15–20 days). Regulation of intake is only possible if dietary energy levels are between 2200 and 3200 kcal DE/kg.

It has already been mentioned that, in order for quantitative requirements to be met, the dietary levels of all other nutrients must be increased if a feed with a high dietary energy concentration, which would be associated with a lower intake, is used.

According to the nature of the basal diet (dietary energy level, concentration and quality of protein) the use of fat may or may not be considered technically worthwhile. In practice, it would seem that its addition is not justified economically.

Within diets currently utilized in France, the low digestibility (10–30%) of cell-wall constituents associated with raw materials such as lucerne or straw means that such materials only have a minor role in meeting energy requirements when compared with starch.

On the other hand, cell-wall constituents from those plants that are only slightly lignified (usually young plants) have a considerably higher digestibility (30–60%) and their contribution to overall energy requirements may be 10–20% (and up to 30% under the most favourable conditions).

Dietary bulk

Cell-wall constituents play an equally important role in providing bulk to the diet. Although the analytical technique is extremely imprecise, measurement of them is generally satisfactorily achieved by the crude fibre method. To ensure an adequate level of bulk, a dietary crude fibre level of 13–14% appears sufficient for the young growing rabbit. For lactating does a slightly lower level (10–11%) is acceptable. As the function of fibre currently being considered is in the provision of bulk, the greater the digestibility of the cell-wall constituents, the higher their dietary level must be such that a level of 10% indigestible crude fibre is maintained.

The bulkiness of the diet and the physical form in which it is presented will influence the dietary energy value of the remainder of the diet: with bulkier feeds the associated increase in rate of passage of digesta will reduce the digestibility of other energy-yielding ingredients (without influencing apparent protein digestibility). In the same way transit time is more rapid the larger the size of particles high in cell-wall constituents: in rabbits, small particles are those less than 0.1–0.2 mm in diameter, and large ones above 0.5 mm. Diets low in bulk (below 10%) are associated with serious digestive problems (risk of diarrhoea which may be fatal).

Protein and amino acids

The sensitivity of the rabbit to dietary protein quality, for a long time controversial, is now accepted. The most recent work has shown that ten amino acids are essential and that an eleventh, glycine, is semi-essential.

Following what is known with other species it could be considered that tyrosine and cystine might partially replace phenylalanine and methionine respectively. In fact the possibility of replacing one sulphur amino acid by the other has been confirmed although no work has been carried out on phenylalanine and tyrosine.

Amino acid requirements for the rabbit have only been studied in practice with respect to lysine, arginine and sulphur amino acids (methionine and cystine). When expressed as a percentage of the diet, lysine and sulphur amino acid levels are around 0.6% for each one, while that for arginine should be more than 0.8%. There is a considerable difference between these optimum levels and those which may be toxic for lysine and arginine. However, with sulphur amino acids, on the other hand, the margin between optimum levels and those excesses associated with poor performance is small.

Minimum levels of other essential amino acids have simply been calculated from those diets that give satisfactory performance. When dietary protein provides all essential amino acids, dietary crude protein levels need only be 15–16%.

For breeding does, optimum levels of crude protein are between 17 and 18% with the same balance of essential amino acids as those given for the young growing rabbit. Precise requirements for essential amino acids in does are completely unknown. An increase in dietary protein level to 21% will raise milk production but will reduce slightly the number of rabbits weaned. A lowering of dietary protein level from 16 to 13% will reduce weaning weight without any appreciable alteration in prolificacy.

In addition, it is known that the feed intake of a diet balanced with respect to essential amino acids is always higher than that with a similar but unbalanced diet. If overall the quality of dietary protein is insufficient, daily dry matter intake is correspondingly reduced which would aggravate the problem.

Finally, with all available information, only protein nitrogen can be used in diets. All preliminary studies investigating non-protein nitrogen sources (urea, ammonium salts) have proved unsuccessful.

Minerals and vitamins

Calcium and phosphorus requirements of growing rabbits are considerably lower than those of lactating does. The latter lose significant amounts of minerals through their milk: 7–8 g daily at peak lactating of which 1.5–2 g is calcium.

An imbalance in dietary levels of sodium, potassium and chloride may promote nephritis and reproductive problems, a risk which is particularly important with those plants, especially lucerne, that are grown with high levels of potassium-containing fertilizers. An improvement in performance has occasionally been observed with dietary levels of copper (200 mg/kg) far higher than those needed to meet requirements. Thus, as with the pig, this could be considered during growth but such higher levels would be more than legally permitted (*see Table 2*).

Microflora within the digestive tract synthesize important amounts of water-soluble vitamins which the rabbit is able to utilize through caecotrophy. In this way, requirements of all B-group vitamins and vitamin C for maintenance and for an average level of production may be met. On the other hand very fast growing animals will respond to additions to the diet of vitamins B_1, B_6 (1–2 mg/kg), B_2 (6 mg/kg) and nicotinic acid (30–60 mg/kg). Dietary levels of up to 1% vitamin C will have no effect positive or negative on the growth and performance of rabbits. Studies on fat-soluble vitamins are less numerous, and required dietary levels have been obtained rather empirically. An excessive level of vitamin D (above 2300 iu/kg diet) is associated with renal and aortic calcification and, therefore, a level of 2000 iu/kg diet should never be exceeded.

Effects of modifications to average diets

The effect of a lack of bulk to the diet is considered on page 65. On the other hand an increase in indigestible crude fibre above 12% will not completely remove the risk of digestive disorders and will worsen food conversion ratio by lowering dietary energy value. When this is below 2200 kcal DE/kg, feed intake capacity is reached, and poorer performance observed with no change in health status of the animal.

Table 24 presents those variations in performance to be expected when levels of dietary protein or certain essential amino acids are lower than those recommended in *Table 23*. These performance levels obtained need not necessarily be economically undesirable as long as dietary protein levels do not fall below 12–13%. This is not the case with minerals (above all calcium and phosphorus) where a deficiency with lactating does would rapidly result.

Table 24 Consequences of a reduction in recommended dietary levels of protein (1%) or certain amino acids (0.1%) on the performance of fattening rabbits (4–11 weeks)

Nutrient under consideration	Reduction in daily liveweight gain		Increase in food conversion ratio		Dietary level below which these relationships will not hold (%)
	Absolute amounts (g/d)	%	Absolute amounts	%	
Crude protein (1%)	−3	−8.5	+0.1	+3	12
Methionine (0.1%)	−2	−6	+0.1	+3	0.4
Lysine (0.1%)	−5	−14	+0.1	+3	0.4
Arginine (0.1%)	−1.5	−4.5	+0.1	+3	0.5

It is difficult to predict a response in the animal following multiple deficiencies. It is advisable to carry out experiments directly to measure the effects arising from the use of available diets if such a situation cannot be avoided. The values therefore proposed in *Table 23*, if used to calculate diet compositions, would ensure that the requirements of the animal are met.

Practical feeding

Presentation of diets

Under controlled conditions of rearing, both dried and ground raw materials are used for rabbits and this allows formulation of balanced compound diets. Unfortunately rabbits do not tolerate dust that is inevitably associated with meal, and therefore it is preferable to use pellet diets.

Table 25 Influence of pellet size on the performance of growing/finishing rabbits (5–12 weeks)

	Pellet diameter (mm)		
	2.5	5	7
Daily feed intake (g)	117	122	131*
Daily.liveweight gain (g)	32.4	33.7	32.0
Food conversion ratio	3.7	3.7	4.1

*An apparent increase in food consumption was in fact due to feed wastage.

For diets currently used, the ideal pellet size is between 3 and 4 mm diameter: it is important never to go beyond 5 mm if wastage is to be avoided (*Table 25*). In order for the pellet to be acceptable to the rabbit it must be less than 8–10 mm in length. Finally, it may be noted that pelleting improves the nutritive value of a diet by between 5 and 7%.

It is possible with certain formulations to feed rabbits meal diets (*Table 26*). A fine particle size which would promote nasal problems should be avoided at all costs. In addition meal diets should not be fed if rabbits are watered from open drinkers which contain water permanently, otherwise the water would become polluted within a few hours and rabbits would then immediately stop drinking and eating. Automatic valve-controlled watering systems must therefore be used. Finally, feeding trials have shown that wet feeds may be used (meal:water in the ratio 60:40) as long as the feeders are kept scrupulously clean.

Table 26 The effect of form of presentation of feed on the performance of young growing rabbits

Reference	Form of presentation	Daily feed intake (g DM)	Daily liveweight gain (g)	Food conversion ratio (feed on DM basis)
Lebas (1973)*	Meal	82	29.7	2.78
	Pellet	94	36.0	2.62
King (1974)**	Meal	79	20.7	3.80
	Pellet	85	22.9	3.70
Machin *et al.* (1980)†	Meal	102	26.5	3.80
	Mash (40% water)	78	27.9	3.06
	Pellet	104	33.1	3.30

Formulation of diets:
*Maize 55.8%, soya-bean meal 25%, barley straw 15%, dl-methionine 0.2%, minerals/vitamins 4%.
**Fish meal 10%, grass meal 20%, wheat bran 40%, oats 12.5%, wheatings 17.5%. Molasses at 1.5% was then added to pellets.
†Barley 62%, soya - bean meal 17.5%, barley straw 12.8%, molasses 5%, lysine 0.25%, methionine 0.05%, minerals 0.93%. The trial was conducted at 25°C.

Choice of feed and level of intake

The traditional method of feeding rabbits was to rely on cereals, bran and forages which were fresh during the summer and dried during the winter. During this latter period, farmers also used fodder beet and carrots. Currently this means of feeding is rapidly declining but it is still used for up to half the rabbits produced in France.

On intensive farms, which represent the major type of commercial production, rabbits are fed compound balanced diets. One feed only is generally used for all ages which corresponds therefore to the mixed diet indicated in *Table 23*.

With a rapid reproductive rate all rabbits, with the exception of males, are fed on an *ad libitum* basis. When the rate is less intense, does are fed the same feed but on a restricted basis from weaning until the end of the subsequent pregnancy. The level of feeding is usually 30–35 g of DM/kg bodyweight per day.

Young growing rabbits are invariably fed on an *ad libitum* basis. When they are group-fed, a single watering point is sufficient for between 10 and 15 individuals: its action must be regularly checked to ensure that the supply is not interrupted even for a short period of time. Similarly only one feeder is sufficient for ten individuals although two are often used to reduce any problems that may occur if one of them were to become blocked. A length of approximately 7–8 cm for each feeder is recommended.

The following values allow the calculation of daily food intake by a group of rabbits:

(1) young growing (4–12 weeks): 110–130 g/day;
(2) lactating does with their litter (weaning at four weeks): 350–380 g/day;
(3) adult at maintenance: 120 g/day;
(4) for the entire farm: 1–1.4 kg/day per doe plus offspring.

Overall, the consumption of pelleted feed on the better French rabbit farms is of the order of 4 kg/kg liveweight sold, including that consumed by the breeding animals. The best figures are 3.6 kg/kg liveweight of 5.9–6.7 kg/kg rabbit meal: this represents the production of 190–220 g of animal protein from 1 kg of plant protein, a return of 19–22%.

Chapter 9
Nutrition of rapidly growing broilers

France produces different types of broiler which are characterized by their age and weight at slaughter (*Table 27*). In this chapter, only recommendations for the three main types, being poussin, small and medium broiler, will be given.

Table 27 Production characteristics of different types of broiler chicken raised in France (1983)

Product	*Age* (d)	*Average liveweight* (g)	*Liveweight range* (g)
Poussin	25–30*	900	800–1000
Small broiler	35–40	1400	1200–1600
Medium broiler	47	1800	1500–2000
Finished broiler	50–54	2200	2000–2400
Roaster	From 81**	2000	1500–2500

*Based on genotype.
**Production unique to France.

Guidelines are given in *Table 28* for liveweight as well as cumulative feed intake and feed conversion ratios for males and females, and on an as-hatched basis, as influenced by age. Non-cumulative weekly consumption of feed and water is presented in *Table 29*.

Presentation of diets and choice of energy level

Body weight gain in broilers is more rapid the higher the daily level of metabolizable energy intake. This level is obviously dependent upon the requirements of the bird, but also upon the form of the diet and its dietary energy level.

Form of the diet

The best liveweights and food conversion ratios are obtained when broilers are fed crumbs during the starter phase followed by pellets (3.5–5 mm diameter). This improvement in performance with pelleting is however only apparent at relatively low dietary energy levels and is barely evident at levels higher than 3200 kcal ME/kg.

Table 28 Performance characteristics of broilers*

	Age (weeks)								
	2	3	4	5	6	7	8	9	10
Males:									
Liveweight (g)	310	580	950	1350	1750	2150	2500	2760	3000
Cumulative feed intake (g)	355	780	1450	2200	3110	4100	5160	6150	7135
Cumulative feed conversion ratio	1.32	1.44	1.59	1.68	1.82	1.94	2.10	2.26	2.41
Females:									
Liveweight (g)	280	530	820	1200	1510	1850	2140	2360	2520
Cumulative feed intake (g)	350	750	1310	2050	2830	3620	4600	5550	6420
Cumulative feed conversion ratio	1.48	1.55	1.68	1.77	1.93	2.00	2.19	2.35	2.55
As hatched:									
Liveweight (g)	295	555	885	1275	1630	2000	2320	2560	2760
Cumulative feed intake (g)	352	765	1380	2125	2970	3860	4880	5850	6780
Cumulative feed conversion ratio	1.38	1.49	1.63	1.72	1.87	1.97	2.14	2.29	2.46

*ME of diet 3100 kcal/kg, ambient temperature 20°C.

Table 29 Weekly feed and water consumption (g) of broilers

Weeks	Males		Females		As hatched	
	Feed	Water	Feed	Water	Feed	Water
1	120	200	120	200	120	200
2	235	375	230	365	232	370
3	425	640	400	600	410	620
4	670	975	560	810	615	890
5	750	1090	730	1050	740	1070
6	910	1395	780	1130	845	1265
7	990	1435	790	1150	890	1292
8	1060	1530	980	1420	1020	1475
9	990	1430	950	1380	970	1405
10	985	1420	870	1260	930	1340

Choice of dietary energy levels

An increase in dietary energy level is invariably associated with an improvement in feed conversion ratio. The effect on growth rate which is influenced by genotype, is apparent up to 3200 kcal ME/kg with birds from 0 to 4 weeks, and up to 3000 kcal ME/kg with those from 4 to 8 weeks of age. Below these values there is a reduction in liveweight at 56 days of around 30 g for each 100 kcal ME/kg reduction in dietary energy level.

Other technological and economic constraints are important when fixing dietary energy levels, particularly:

(1) Problems associated with the manufacture, handling and storage of diets high in fat.
(2) Fat content of carcasses if the slaughter age is beyond six weeks. Within the usual range of dietary energy levels (2800–3200 kcal ME/kg) each increase of 100 kcal/kg from four weeks of age is associated with an additional fat content of 2% of carcass weight.
(3) The increase in the unit cost of dietary energy at higher levels. As the energy intake of broilers is practically the same whether the dietary level is equal to or higher than that required to promote maximum growth rate, the dietary energy level selected must be that which is least costly.

The establishing of a dietary level by the feed formulator needs to account for all these points (price of raw materials, slaughter age, genotype, degree of fattening required, etc.). In all cases, each measurement of bird performance, and particularly food conversion ratio, has only a relative value based essentially on the economic context in which it was established.

Protein and amino acids

In order to simplify and clarify the situation, recommended levels of amino acids are expressed as a function of dietary energy levels. However it is known that, in addition to dietary energy concentration, ambient temperature, density of the feed and other factors may modify energy intake of broilers which would be associated with a corresponding alteration in amino acid intake. Thus the recommendations provided are only relevant at both the growth rates and feed intakes presented in *Table 28*. In situations of under- or overconsumption, it is therefore necessary to modify the suggested figures for amino acids given in *Tables 30* to *34* to obtain the same level of performance.

Within these tables the use of diets with 2900, 3000, 3100 or 3200 kcal ME/kg has been anticipated. For each of these dietary energy levels, the minimum amounts of amino acids necesary for the growth rates in *Table 28* are given taking into account the age and sex of the bird.

Excess levels of amino acids will not reduce performance as long as certain imbalances are avoided. In the case of excess lysine (very infrequently encountered), the balance with arginine must be maintained and it must never be less than 0.7 of the level of lysine. Similarly relative levels of leucine, isoleucine and valine have to be kept constant: the balance between that amino acid in excess and each of the other two should never differ by more than 1.5 times that calculated from *Tables 30* to *34*.

For crude protein (N × 6.25) it is possible to choose between two levels, being minimum and optimum, which more or less promote the same growth rate but not the same efficiency of food utilization: food conversion ratio resulting from the minimum is slightly inferior.

As a guideline it may be said that a reduction in dietary protein level of 1% has the same effect on food conversion ratio as a fall in dietary energy level of 50 kcal ME/kg. However, as before, bird performance is not the only criterion that needs to be considered. Taking into account the price of protein sources, dietary regimes with minimum levels are generally the cheapest.

Table 30 Recommended dietary energy levels and concentrations of protein, amino acids and minerals for the starter poussin (males and females) during the first two weeks*

	Dietary energy level (kcal ME/kg)			
	2900	3000	3100	3200
Crude protein (%)	21.5	22.2	23.0	23.7
Amino acids (%):				
Lysine	1.12	1,16	1.20	1.24
Methionine	0.47	0.48	0.50	0.52
Sulphur amino acids	0.84	0.87	0.90	0.93
Tryptophan	0.20	0.21	0.22	0.23
Threonine	0.67	0.70	0.72	0.74
Glycine + serine	1.87	1.94	2.00	2.06
Leucine	1.57	1.63	1.68	1.73
Isoleucine	0.89	0.92	0.95	0.96
Valine	0.98	1.01	1.04	1.08
Histidine	0.45	0.46	0.48	0.50
Arginine	1.21	1.26	1.30	1.34
Phenylalanine + tyrosine	1.50	1.55	1.60	1.65
Minerals (%):				
Calcium	1.00	1.03	1.06	1.10
Total phosphorus	0.67	0.68	0.69	0.70
Available phosphorus	0.42	0.43	0.44	0.45
Sodium**	0.16	0.16	0.17	0.17
Chloride**	0.14	0.14	0.15	0.15

*Only for the first week if three successive diets are used. The exact time of changeover will vary by a few days depending upon the type of production.
**Wherever possible, a balance between sodium, potassium and chloride should be maintained at (Na + K)−CI = 250 mmol/kg or meq/kg.

Finally, it must be appreciated that optimum levels are associated with leaner broilers: the level of carcass fat falls by around 0.35% with an increase in dietary protein level of 10 g/kg beyond the minimum.

Characteristics of each production system

Poussins

Occasionally, poussins are selected, and subsequently slaughtered, from the largest birds in the flock. In the case of specialist farms, it is preferable to use only males and to use a starter followed by a grower feed (*Tables 30* and *31*). The time of changeover of diets and the age of slaughter is dependent upon genotype (broiler or layer-type).

Small broilers

It is recommended that two diets be used: a starter replaced by, between 14 and 21 days, a grower with a minimum or optimum protein level (*Table 31*).

Table 31 Recommended dietary energy levels and concentrations of protein, amino acids and minerals for the growing broiler (as hatched or males) during the third week*

	Dietary energy level (kcal ME/kg)			
	2900	3000	3100	3200
Crude protein (%)†	19.6 (16.8)	20.4 (17.4)	21.0 (18.0)	21.7 (18.6)
Amino acids (%):				
Lysine†	0.98 (0.88)	1.02 (0.92)	1.05 (0.95)	1.08 (0.98)
Methionine†	0.43 (0.36)	0.44 (0.37)	0.46 (0.38)	0.47 (0.39)
Sulphur amino acids†	0.75 (0.68)	0.77 (0.70)	0.80 (0.72)	0.83 (0.74)
Tryptophan	0.19	0.20	0.21	0.22
Threonine	0.59	0.61	0.63	0.65
Glycine + serine	1.64	1.69	1.75	1.81
Leucine	1.38	1.42	1.47	1.52
Isoleucine	0.78	0.80	0.83	0.86
Valine	0.86	0.89	0.92	0.95
Histidine	0.39	0.41	0.42	0.43
Arginine	1.03	1.06	1.10	1.14
Phenylalanine + tyrosine	1.31	1.35	1.40	1.45
Minerals (%):				
Calcium	0.90	0.93	0.97	1.00
Total phosphorus	0.66	0.67	0.68	0.69
Available phosphorus	0.41	0.42	0.43	0.44
Sodium**	0.16	0.16	0.17	0.17
Chloride**	0.14	0.14	0.15	0.15

*See footnote to *Table 30*
**See footnote to *Table 30*.
†The figures given in brackets correspond to requirements when growth only is considered.

Medium broilers

Whatever the price of raw materials, the use of only two diets (from 0 to 3 and from 4 to 7 weeks) is associated with a waste of protein. Accordingly, successive use of starter (0 to 7 days), grower (8 to 21 days) and finisher (after 21 days) diets is recommended. The price of protein-rich raw materials and the anticipated growth rate will determine as before the choice between optimum and minimum protein levels.

Sexed birds

The rearing of birds separately according to sex may allow significant savings in protein and is therefore of greater interest if its price is high. Beyond 15 days of age, protein requirements of females are in fact lower than those of males and become increasingly so with age: special diets for females must therefore be fed during the grower and finisher periods (*Tables 32* and *34*). As females have a greater potential for fat deposition, a lower dietary energy level (maximum of 3000 kcal ME/kg) than that used for males must be employed.

Table 32 Recommended dietary energy levels and concentrations of protein, amino acids and minerals for the growing broiler (female) from the third week*

	Dietary energy level (kcal ME/kg)		
	2900	3000	3100
Crude protein (%)†	18.7 (16.4)	19.4 (16.9)	20.0 (17.5)
Amino acids (%):			
Lysine†	0.94 (0.85)	0.97 (0.88)	1.00 (0.91)
Methionine†	0.41 (0.35)	0.43 (0.36)	0.44 (0.37)
Sulphur amino acids†	0.71 (0.66)	0.74 (0.68)	0.76 (0.70)
Tryptophan	0.19	0.19	0.20
Threonine	0.56	0.58	0.60
Glycine + serine	1.56	1.62	1.67
Leucine	1.31	1.35	1.40
Isoleucine	0.74	0.76	0.79
Valine	0.82	0.85	0.88
Histidine	0.37	0.39	0.40
Arginine	0.98	1.02	1.05
Phenylalanine + tyrosine	1.24	1.24	1.33
Minerals (%):			
Calcium	0.90	0.93	0.97
Total phosphorus	0.63	0.64	0.65
Available phosphorus	0.38	0.39	0.40
Sodium**	0.16	0.16	0.17
Chloride**	0.14	0.14	0.15

*, **, † See footnotes to Tables 30 and 31.

Minerals, vitamins and additives

The recommendations collected together in *Tables 30* to *34* deal with levels of calcium and phosphorus: these are calculated to cover requirements precisely. Higher levels of phosphorus may improve performance slightly. For example a level of 0.45% of available phosphorus during finishing compared with 0.35% will increase liveweight by 40 g and improve food conversion ratio by around 0.05: the choice of which to use is one of economics. Trace element additions (*Table 35*) include a large margin of security to account for variations in the composition of raw materials.

Table 35 contains, additionally, recommended added levels of vitamins (and not total dietary levels). The use of excesses is justified in part by their low cost and also by knowledge of subsequent deficiencies which are always possible (raw material quality, problems of homogenization, stability within the feed). These high levels, which are usually employed without problems, must however be restricted for vitamins A and D (which should not be above 3.5 times the values indicated in *Table 35*).

Table 33 Recommended dietary energy levels and concentrations of protein, amino acids and minerals for the finishing broiler (as hatched or male) from three weeks

	Dietary energy level (kcal ME/kg)			
	2900	3000	3100	3200
Crude protein (%)†	18.2 (14.8)	18.9 (15.3)	19.5 (15.8)	20.1 (16.3)
Amino acids (%):				
Lysine†	0.84 (0.74)	0.87 (0.77)	0.90 (0.80)	0.93 (0.83)
Methionine†	0.38 (0.30)	0.39 (0.31)	0.40 (0.32)	0.41 (0.33)
Sulphur amino acids†	0.69 (0.61)	0.71 (0.63)	0.73 (0.65)	0.75 (0.67)
Tryptophan	0.16	0.16	0.17	0.18
Threonine	0.48	0.49	0.51	0.53
Glycine + serine	1.33	1.37	1.42	1.47
Leucine	1.11	1.15	1.19	1.23
Isoleucine	0.63	0.65	0.67	0.69
Valine	0.55	0.57	0.59	0.61
Histidine	0.32	0.33	0.34	0.35
Arginine	0.86	0.89	0.92	0.95
Phenylalanine + tyrosine	1.06	1.09	1.13	1.17
Minerals (%):				
Calcium	0.80	0.83	0.87	0.90
Total phosphorus	0.60	0.61	0.62	0.64
Available phosphorus	0.35	0.36	0.37	0.38
Sodium**	0.16	0.16	0.17	0.17
Chloride**	0.14	0.14	0.15	0.15

**,† See footnotes to Tables 30 and 31.

Table 34 Recommended dietary energy levels and concentrations of protein, amino acids and minerals for the finishing broiler (female) from three weeks

	Dietary energy level (kcal ME/kg)		
	2900	3000	3100
Crude protein (%)†	18.2 (13.1)	18.9 (13.5)	19.5 (14.0)
Amino acids (%):			
Lysine†	0.74 (0.66)	0.77 (0.68)	0.80 (0.70)
Methionine†	0.34 (0.28)	0.35 (0.29)	0.36 (0.30)
Sulphur amino acids†	0.61 (0.53)	0.63 (0.55)	0.65 (0.57)
Tryptophan	0.14	0.15	0.15
Threonine	0.42	0.44	0.45
Glycine + serine	1.17	1.21	1.25
Leucine	0.98	1.02	1.05
Isoleucine	0.55	0.57	0.59
Valine	0.50	0.51	0.53
Histidine	0.28	0.29	0.30
Arginine	0.76	0.78	0.81
Phenylalanine + tyrosine	0.93	0.97	1.00
Minerals (%):			
Calcium	0.80	0.83	0.87
Total phosphorus	0.57	0.58	0.59
Available phosphorus	0.32	0.33	0.34
Sodium**	0.16	0.16	0.17
Chloride**	0.16	0.16	0.17

**, † See footnotes to Tables 30 and 31.

Table 35 Recommended added amounts of trace elements and vitamins for the broiler (mg/kg or iu)

	Starter/grower	Finisher
Trace elements:		
Iron	40	15
Copper	3	2
Zinc	40	20
Manganese	70	60
Cobalt	0.2	0.2
Selemium	0.1	0.1
Iodine	1	1
Vitamins:		
Vitamin A (iu)	10 000	10 000
Vitamin D_3 (iu)	1 500	1 500
Vitamin E	15	10
Vitamin K_3	5	4
Thiamine	0.5	–
Riboflavin	4	4
Pantothenic acid	5	5
Niacin	25	15
Folic acid	0.2	–
Vitamin B_{12}	0.01	0.01
Choline chloride	500	500

The action of some coccidiostats is based upon their antivitamin (B_1 or folic acid) activity. The result of an excess of B_1 or folic acid may be, partly, to neutralize the protective function of the coccidiostat and to produce symptoms of the disease, either acute or sub-clinical (a simple lack of pigmentation in the case of yellow broilers).

The recommended level of vitamin E, in *Table 35*, is sufficient on condition that the legally permitted amount of an antioxidant is also added.

Broilers reared on litter must be fed diets containing approved coccidiostats at permitted levels. The use of antibiotics and other growth factors may slightly improve performance, particularly if rearing conditions are sub-optimal.

To produce yellow broilers, the particular diet used (based on maize, gluten, lucerne) may be supplemented with synthetic carotenoids, or plant products rich in xanthophylls. In the latter case it is important to ensure that the pigments are actually active, as carotenoids are frequently altered during extraction. The intensity of colour produced is dependent upon the dietary level of pigments: a level of 20–30 ppm is sufficient for satisfactory colour and it is possible within this zone of incorporation to compare the relative value of different sources of carotenoids. It is around 0.8 for lucerne and other plant sources rich in lutein (marigolds), 1.0 for maize and gluten and 1.4 for the apocarotene ester. In order to obtain more pronounced colour (golden or orange) it is preferable to utilize synthetic carotenoids which are considerably more effective at these higher levels of coloration.

Chapter 10
Nutrition of laying hens

Type of bird considered

Commercial laying hens used in egg production are of two types which differ in many respects (egg-shell colour, adult liveweight). The average performance (established during 1978/79 at various European testing stations) of these two types is given in *Table 36*.

Table 36 Production characteristics of pullets and laying hens

	White layers (Leghorn)	Brown layers (Rhode Island)
Liveweight (kg):		
At 20 weeks	1.3	1.6
At 70 weeks	1.6	2.2
Age (d) at 50% lay	159	159
Eggs laid by 70 weeks of age	269	264
Mean egg weight (g)	60.6	63.0
Feed intake (kg/bird)*:		
0–20 weeks	6.6	7.6
21–70 weeks	40.0	45.7
Food conversion ratio		
(kg feed/kg eggs)	2.45	2.75
Mortality (%):		
0–20 weeks	3.8	1.5
21–70 weeks	6.8	3.0

*Feed of 2800 kcal ME/kg at an ambient temperature of 17°C.

Broiler breeders originated from White Rock (females) and Cornish (males) breeds. Certain breeder hens, characterized by the dwarfing gene *dw*, are smaller in size and consequently have modified nutrient requirements. Average performance of normal and dwarf hens are presented in *Table 37*.

Thus four types of laying hen may be defined:

(1) Laying hens for egg consumption (called layers):
   (a) white layers;
   (b) brown layers.

Table 37 Production characteristics of broiler breeder hens

	Normal hen	Dwarf hen	Cock
Liveweight (kg):			
At 22 weeks	1.95	1.75	2.70–3.00
At 65 weeks	3.20	2.60	5.0
Age (d) at 50% lay	196	190	
Eggs laid by 65 weeks of age:			
Total	157	164	
Viable	150	158	
Chicks hatched per hen	127	136	
Mean egg weight (g)	63.5	62.0	
Feed intake (kg/bird)*:			
0–24 weeks	10	8.7	11.3
25–65 weeks	41	34	51
Feed intake (g) per viable egg**:			
Rearing + lay + cock	382	315	
Lay + cock	307	247	
Mortality and culling (%):			
0–24 weeks	2	2	30
25–65 weeks	4	6.8	20

*Feed of 2800 kcal ME/kg at an ambient temperature of 17°C.
**Appropriate for a flock where the ratio of cocks to hens is 1:10.

(2) Broiler breeder hens (called breeders):
  (a) normal;
  (b) dwarf.

The relative nutrient recommendations of breeders are not considered subsequently and are identical to those for current types of layer with the exception of trace elements and vitamins, which must always be higher when eggs produced are to be hatched.

Nutrition of rearing pullets

Generally, nutrition during rearing has little influence upon subsequent laying performance. It is therefore not necessary to promote a rapid growth rate; rather, it is important to attain sexual maturity at a given age and liveweight with the minimum of nutrient costs.

The use of lighting regimes constitutes the most effective means of controlling point of lay. However, a deficiency of protein or an overall feed restriction may also, to a lesser degree, delay point of lay: under the most severe conditions, this effect of diet may be of the order of two weeks. The use of non-digestible dietary diluents (cellulose) is the most inefficient. Of all the methods proposed, that based on the overall restriction of a balanced diet is the most useful and economic.

In practice, diets of average energy level (2600–2800 kcal ME/kg) are used which correspond to the lowest unit energy cost. Other dietary constituents are provided at those levels, indicated in *Table 38*, which are just sufficient. Necessary quantities of minerals and vitamins are those already established for the growing broiler (cf. Chapter 9).

Table 38 Recommended dietary energy levels and concentrations of protein, amino acids and minerals for the pullet

	Starter (0–6 weeks)	Rearing (6 weeks until the first egg), feed restricted	Rearing (8 weeks until the first egg), feed ad libitum
Dietary energy level (kcal ME/kg)		Less than 2900	
Crude protein (%) for a dietary energy level of 2800 kcal ME*	18.0	14.5	13.0
Lysine (%)	0.85	0.65	0.55
Methionine (%)	0.33	0.28	0.26
Sulphur amino acids (%)	0.65	0.50	0.46
Minerals and vitamins	As for the female broiler over 2 weeks of age (see Table 32)**		
Coccidiostats	Maximum recommended levels†		Recommended levels†

*For different dietary energy levels, the required protein levels may be calculated by multiplying the given values by:

$$\frac{\text{new ME level in kcal/kg}}{2800}$$

**With the exception of available phosphorus, which may be lowered to a dietary level of 0.3% from eight weeks of age when feed is restricted.
†Within legal limits.

In general two diets are sufficient from hatching until point of lay: a starter feed for the first six to eight weeks and a grower feed thereafter. In the absence of any quantitative feed restriction, a grower feed with the protein and amino acid levels corresponding to those in the last column of *Table 38* may be used. With feed restriction, the values in the middle column may be adopted.

Feed restriction must be imposed very early (from five weeks of age) and with an increasing degree of severity as indicated in *Table 39*. Although not essential for white layers, it is recommended for brown layers and broiler breeders. From a nutritional point of view, daily restriction is preferable; if rearing conditions do not properly allow this (insufficient feeder length, chains too slow) twice the daily amount may be distributed every other day (skip-a-day method). The programmes of feeding outlined are calculated to give a point of lay of 21 and 24 weeks of age respectively for brown layers and dwarf breeders. If an earlier point of lay is required, the degree of restriction imposed must be less severe, particularly after 15 and 17 weeks of age respectively for the two types, and of the order of 10% above the levels indicated.

Coccidiostats must be used at the maximum permitted levels when feed is restricted. To avoid the build-up of resistance, it is recommended that different products be used for starter and grower feeds.

Restriction must only be gradually relaxed from point of lay and stopped by the time that rate of lay of the flock is 25%. If this is undertaken too quickly, then overconsumption will occur and the benefits of previous feed restriction will have been lost: in addition it may lead to pathological problems including liver haemorrhage.

Table 39 Daily feed allowances for layers and broiler breeders during rearing (quantities in g/bird per day)

Age (weeks)	White layers	Brown layers	Heavy broiler breeder	Dwarf broiler breeder	Heavy cock broiler
0–5	ad libitum	ad libitum	ad libitum	ad libitum	ad libitum
6	–	48	50	40	50
7	–	50	55	45	55
8	–	55	60	47	60
9 and 10	–	60	65	49	64
11 and 12	–	65	65	52	73
13 and 14	–	70	70	55	83
15 and 16	–	75	70	56	93
17 and 18	–	80	75	58	98
19 and 20	–	90	75	62	110†
21	–	100	85	67	120
22	–	105	90	70	130
23	–	ad libitum at a rate of lay of 25%	100	75	Placed with laying hens
24	–		115	85	
25	–		130	100	
26	–		ad libitum at a rate of lay of 25%	ad libitum at a rate of lay of 25%	
27	–				

Values are for feed of 2800 kcal ME/kg* at an ambient temperature of 17°C**.
*For different dietary energy levels, the required amounts may be calculated by multiplying the given quantities by:

$$\frac{2800}{\text{ME level of diet offered in kcal/kg}}$$

**Quantities offered need to be reduced by 1% for each 1°C rise in temperature, or increased with each 1°C fall.
†Amount used for adult cocks kept in cages for artificial insemination.

The feed restriction of cocks outlined in *Table 39* is only possible if they are reared separately: if properly undertaken, it promotes excellent fertility and a reduced level of adult mortality.

In all cases the amount of food given must be adjusted as a function of dietary energy level and ambient temperature. The values in *Table 39* correspond to a diet of 2800 kcal ME/kg at an ambient temperature of 17°C; the corrections necessary if these conditions change are indicated.

Nutrition of hens during lay

The laying diet gradually replaces the rearing diet from the moment that the first egg appears (being about two weeks before 50% lay). It must be offered on an *ad libitum* basis for the first few months of lay from the point at which rate of lay is 25%.

Energy requirements

Energy requirements of laying hens are dependent above all on their liveweight (maintenance) but also to a degree on liveweight increase, feathering and rate of lay. The equations allowing requirements to be properly met are given on page 13.

The importance of feed consumption in the fulfilling of energy requirements is paramount with white layers but less so with brown layers, which tend to overconsume energy with diets of higher energy concentration at heavier liveweights. Other than for Leghorn types (white layers) it is preferable to use diets of moderate energy concentration (2500–2800 kcal ME/kg).

The influence of temperature is important and concerns only maintenance requirements. With layers, these are reduced by 4 kcal/day for each 1°C increase in temperature from 0 to 29°C. With heavy breeders, the fall is of the order of 6 kcal/day per 1°C. Beyond 30°C maintenance requirements become considerably lower; a reduction in feed intake is observed associated with poorer performance.

Protein and amino acid requirements

Protein requirements are influenced far more by level of egg production (number and average size of eggs) than by liveweight of the bird. Maintenance of liveweight, whatever the type, requires only 2–4 g of protein daily whereas egg formation needs 10–12 g. At peak lay, heavy and light genotypes therefore have more or less the same requirements and it is possible to define, whatever the type, the minimum daily amounts of amino acids necessary to promote maximum production.

Table 40 Daily energy and nutrient allowances for laying hens at the start of lay. Quantities (in g/day) are designed to promote maximal egg production and shell quality

Energy	Variable influenced by genotype and temperature
Crude protein	16.0
Amino acids:	
Lysine	0.75
Methionine	0.34
Sulphur amino acids	0.61
Tryptophan	0.17
Valine	0.65
Threonine	0.52
Minerals:	
Calcium	4.20
Total phosphorus	0.60
Available phosphorus	0.35
Sodium	0.16
Chloride	0.15
Linoleic acid	1.00

The values presented in *Table 40* have resulted from numerous trials with both white and brown layers. They are also applicable to breeders although some results have indicated that these have a slightly higher daily protein requirement (18 g as opposed to 16 g) as a consequence of the development of the egg-laying tract which occurs after the period of feed restriction imposed during rearing. The protein requirement is higher the earlier the point of lay. As a consequence, those flocks that come into lay early (20 weeks for layers and 22 weeks for dwarf breeders) need a diet higher in protein.

As a general rule it is advisable, taking into account raw material variability, to provide a slightly higher level of dietary protein than that necessary to meet requirements as this will remove risks of deficiency. Above all, whatever the origin (genetic or physiological) an unusually heterogeneous flock will artificially increase mean requirements. Finally feed consumption may be modified by temperature. The recommendations presented in *Table 41* account for all these factors and also include a margin of security. In addition, characteristics of a diet appropriate to hot climates are given. The quantity of protein allowed in diets for heavy breeders is often very high; therefore any additional dietary amount would constitute wastage as the requirement for reproduction is not higher than that for lay.

In the event of a shortage in protein sources, the values in *Table 40* may be used as a basis for the direct formulation of feeds. They would evidently result in low dietary protein levels, but excellent levels of performance have been obtained with layers and breeders fed diets with only 12–13% protein: egg numbers are unchanged, and only a small reduction in egg size results (around 1 g) with, frequently, a slight worsening in food conversion ratio. The important point is that, in this case, it is essential to monitor amino acid balance as it is known with maize-soya mixtures poor in protein, limiting factors may be methionine followed by lysine and, depending upon genotype, tryptophan, threonine or valine.

Table 41 Recommended dietary energy levels and concentrations of protein, amino acids and minerals for the laying hen and broiler breeder

	Layers and dwarf broiler breeders*		Heavy broiler breeders		Layers in hot climates
Dietary energy level (kcal ME/kg):	2600	2800	2600	2800	2800
Crude protein (%)	14.0	15.0	12.0	13.0	18.5
Amino acids (%):					
Lysine	0.63	0.68	0.51	0.55	0.93
Methionine	0.28	0.30	0.24	0.26	0.41
Minerals (%):					
Calcium	3.4**	3.6**	2.8	3.0	4.0
Total phosphorus	0.56	0.58	0.53	0.56	0.65
Available phosphorus	0.31	0.33	0.28	0.31	0.40
Sodium	0.13	0.14	0.10	0.12	0.15
Chloride	0.13	0.14	0.10	0.12	0.15
Linoleic acid (%)	0.8	0.9	0.6	0.7	1.0
Xanthophyll (mg/kg)	23	25			30
Expected daily feed intake (g) at 18°C	127	120	170	160	

*For Leghorns kept at an ambient temperature above 25°C, a diet intermediary between that for layers/dwarf breeders and layers in hot climates (last column) is recommended.
**In the case of dwarf broiler breeders, a level in excess of 3.2% is not recommended.

Minerals and vitamins

Phosphorus requirements of laying hens are low. A relatively high supplementary level (*Tables 40* and *41*) is, however, recommended particularly to take into account problems of homogenization of diets. High levels of calcium (above 3.4%) are in all cases essential to ensure adequate shell strength. Towards the end of lay, during hot conditions and under other circumstances where egg-shell strength deteriorates, it is possible to replace 50–60% of the dietary calcium carbonate in powdered form by a specific source of calcium (oyster shells, egg shells, carbonate granules) that allows the hen to consume calcium independently of other nutrients.

Dietary chloride levels must be limited to 0.14% which is equivalent to 0.23% of sodium chloride. In the absence of this form, sodium may be additionally provided as bicarbonate, carbonate or sulphate on condition that the latter is not at a level higher than 0.25%. Trace element and vitamin additions are detailed in *Table 42*. For breeders, vitamin levels are higher to ensure hatchability; requirements for reproduction are in effect frequently higher than those for lay.

Table 42 Recommended added dietary levels for trace elements and vitamins for laying hens and laying broiler breeders (mg/kg or iu)

Trace elements for all genotypes and climates	
Iron	40.0
Copper	2.0
Zinc	40.0
Manganese	60.0
Cobalt	0.2
Selenium	0.15
Iodine	0.8

Vitamins for all climates	Laying hens	Broiler breeders (heavy and dwarf)
Vitamin A (iu)	8 000	10 000
Vitamin D_3 (iu)	1 000	1 500
Vitamin E	5	15
Vitamin K_3	2	4
Riboflavin	4	4
Calcium pantothenate	4	8
Pyridoxine	0	1
Biotin	0	0.1
Folic acid	0	0.2
Vitamin B_{12}	0.004	0.008
Choline chloride	250	500

Feed restriction during lay

Layers

Although restriction during rearing has only a very minor influence on subsequent laying performance, it has a considerable effect if imposed during lay. The margin which separates savings that arise from this reduction is narrow and any feed restriction, however small, results in a reduction in egg number whereas the effect on mean egg weight appears less evident. However, some commercial layer strains have a tendency to overconsume and a reduction through a degree of feed restriction undertaken cautiously may be of benefit; it results in effect in feed savings and occasionally a longer period of lay.

It seems that a system of feed restriction is inappropriate for the Leghorn. On the other hand, brown layers may be restricted in moderation from the fourth month of lay (95% of *ad libitum* intake). This may be achieved either by the distribution of the given amount of feed, or by limiting time of access to feeders (four hours per day); the time is dependent upon genotype and the form of the diet (meal or crumbs) and must be adjusted as a function of the level of consumption hoped for. In all cases, distribution of feed in the afternoon is essential in particular for egg-shell quality. The judicious use of intermittent lighting regimes (many light-dark cycles during 24 hours) may also result in appreciable feed savings often associated with other effects (slight reduction in rate of lay compensated for by an increase in egg size and shell quality).

Breeders

Breeders must be restricted after the fourth month of lay. Competent farmers may impose restriction much earlier and throughout lay after peak production. The amount of feed offered as a function of ambient temperature is given in *Table 43*. These values are only provided as guidelines and they must be adapted according to the commercial strain used and the rearing conditions utilized.

Table 43 Feed allowances for broiler breeder hens from four months into lay*

	Energy allowances (kcal ME/bird per day)			*Daily feed allowances* (g/bird per day) at dietary ME levels of:					
				2600 kcal/kg			*2800 kcal/kg*		
Ambient temperature (°C)	15	20	25	15	20	25	15	20	25
Heavy breeders	425	395	365	163	152	140	152	141	130
Dwarf breeders	330	310	290	127	119	112	118	111	104

*Allowances for cocks need to be added to these figures. These may be calculated as:

1.3 that of heavy breeder hens;
1.5 that of dwarf breeder hens.

Pigmentation of yolk

It is essential to incorporate raw materials with adequate levels of xanthophylls in diets for layers to ensure satisfactory yolk colour. It will be recalled that on average a xanthophyll concentration of 25 mg/kg will produce the required colour (10 on the Roche fan). If a substantial amount of maize is replaced by wheat or barley, it is vital that other sources of pigments are utilized, either naturally occurring (lucerne meal or protein, maize gluten, algae, etc.) or synthetic (apocarotene ester). When yellow pigments are insufficient, the addition of traces of red pigments (1–2 mg/kg of pure canthaxanthine) will considerably intensify the yellow colour by adding a touch of orange appreciated by consumers. *Table 99* represents levels of xanthophylls and their availability in most principal raw materials used in the nutrition of laying hens.

Chapter 11
Nutrition of turkeys

Meat turkeys

Three types of turkey are raised in France; they differ above all in terms of growth rate and also by virtue of feather colour. The farm type has the slowest growth rate and corresponds to the Christmas roasting turkey. The medium and heavy types are each destined for the cutting and jointing industry. The characteristics of these three types are summarized in *Table 44* and they serve as the basis of the nutrient recommendations which follow: *Tables 45* and *46* present weekly water and feed intakes of each type.

Choice of energy level

The turkey is less sensitive than the broiler to the dietary metabolizable energy content which should permit the use of a relatively wide range of energy levels. In practice, however, the turkey frequently has insufficient body fat depots which may be augmented by raising dietary energy levels (in particular by using fat); therefore high levels need to be used, above all during the finisher phase.

Protein and amino acids

The particularly long rearing period of this species justifies the use of a wide range of feeds. Changes carried out every four weeks allow dietary levels to match requirements without complications for the feed formulator or farmer. The recommendations and corresponding time periods are detailed in *Tables 47* to *52*. They presuppose, of course, that health conditions during rearing are satisfactory.

Minerals and vitamins

The low feed consumption of turkeys means that higher dietary levels of minerals and vitamins than those used for the broiler are necessary, especially during the first 12 weeks. Recommended levels of minerals are given in *Tables 47* to *52*; those concerning trace elements and vitamins are presented in *Table 53*.

Breeders during rearing

During rearing, breeders do not need such concentrated diets as do those reared for consumption. In practice it is advisable to use three successive feeds (*Table 54*) during the periods 0–4, 5–14 and 15–25 weeks respectively. Beyond this age, layer

Table 44 Performance characteristics of turkeys*

		_____ Age (weeks) _____						
		4	8	12	14	16	20	24
Farm type**								
Males	Liveweight (g)	600	1950	3350		4700	6050	
	Cumulative feed intake	880	3650	7420		12680	18900	
	Cumulative feed conversion ratio	1.60	1.92	2.25		2.73	3.15	
Females	Liveweight (g)	500	1550	2550		3500	4500	
	Cumulative feed intake	770	2970	6320		10670	15860	
	Cumulative feed conversion ratio	1.70	1.98	2.53		3.09	3.56	
Medium type:								
Males	Liveweight (g)	830	3000	5700	7000	8100		
	Cumulative feed intake	1160	5050	12170	16950	21950		
	Cumulative feed conversion ratio	1.51	1.72	2.16	2.44	2.73		
Females	Liveweight (g)	740	2450	4350				
	Cumulative feed intake	1070	4320	9555				
	Cumulative feed conversion ratio	1.57	1.81	2.23				
Heavy type:								
Males	Liveweight (g)	820	3025	6050	7500	8900	11500	14000
	Cumulative feed intake	1160	5120	12290	22420	22420	34610	46950
	Cumulative feed conversion ratio	1.53	1.73	2.06	2.30	2.54	3.02	3.37
Females	Liveweight (g)	720	2410	4550	5600	6700	8250	
	Cumulative feed intake	1080	4490	10260	13930	17910	26470	
	Cumulative feed conversion ratio	1.64	1.91	2.28	2.51	2.70	3.23	

*At a dietary energy level of 3000 kcal ME/kg and an ambient temperature of 20°C.
**Found in some European countries.

diets may be used if the sexual maturity of the genotype in question permits it.

The dietary energy level of feeds used after four weeks must not be higher than 2900 kcal ME/kg.

A slight degree of feed restriction may be imposed between 12 and 18 weeks following advice from the breeder. All feeds may be in granular form.

Breeders

Egg mass laid by the turkey is smaller than that of the chicken when expressed in terms of body weight and consequently breeder turkeys may be fed on diets of lower amino acid and mineral levels.

Table 45 Weekly feed and water intakes (g) for farm type and medium type turkeys

Weeks	Farm type				Medium type			
	Males		Females		Males		Females	
	Feed	Water	Feed	Water	Feed	Water	Feed	Water
1	60	135	60	135	80	175	80	175
2	160	355	150	355	210	465	200	465
3	270	565	240	505	350	735	310	650
4	390	820	320	670	520	1090	480	1010
5	530	1060	420	840	700	1400	600	1200
6	660	1320	520	1040	875	1750	750	1500
7	750	1500	590	1180	1055	2110	890	1780
8	830	1660	670	1340	1260	2520	1010	2020
9	900	1800	750	1500	1470	2940	1125	2250
10	980	1860	790	1500	1680	3190	1250	2375
11	1060	2015	870	1655	1870	3555	1370	2605
12	1130	2145	940	1785	2100	3990	1460	2830
13	1210	2300	1000	1900	2320	4410		
14	1280	2430	1060	2015	2460	4675		
15	1350	2565	1120	2130	2480	4710		
16	1420	2700	1170	2225	2520	4790		
17	1480	2810	1230	2340				
18	1540	2925	1280	2435				
19	1580	3000	1320	2510				
20	1620	3080	1360	2585				

Table 46 Weekly feed and water intakes (g) for heavy type turkeys

Weeks	Males		Females	
	Feed	Water	Feed	Water
1	80	175	80	175
2	210	465	200	435
3	350	735	310	650
4	520	1090	490	1030
5	710	1420	630	1260
6	890	1780	780	1560
7	1080	2160	930	1860
8	1280	2560	1070	2140
9	1460	2980	1220	2440
10	1690	3210	1370	2605
11	1890	3590	1520	2890
12	2130	4045	1660	3155
13	2340	4445	1790	3400
14	2480	4710	1880	3570
15	2570	4885	1960	3725
16	2740	5205	2020	3840
17	2910	5530	2080	3950
18	3050	5795	2120	4030
19	3120	5930	2160	4105
20	3110	5910	2200	4180
21	3100	5890		
22	3100	5890		
23	3080	5850		
24	3060	5815		

Table 47 Recommended dietary concentrations (%) of protein, amino acids and minerals for turkeys during the starter phase (0–4 weeks)

	Dietary energy level (kcal ME/kg)			
	2800	2900	3000	3100
Crude protein	24.3	25.1	26.0	26.9
Amino acids:				
Lysine	1.64	1.70	1.76	1.82
Methionine	0.44	0.45	0.47	0.49
Sulphur amino acids	1.12	1.16	1.20	1.24
Tryptophan	0.23	0.24	0.25	0.26
Threonine	0.89	0.92	0.95	0.98
Glycine + serine	2.45	2.55	2.65	2.75
Leucine	1.50	1.55	1.60	1.65
Isoleucine	0.79	0.82	0.85	0.88
Valine	1.12	1.16	1.20	1.24
Histidine	0.60	0.62	0.64	0.66
Arginine	1.59	1.64	1.70	1.76
Phenylalanine + tyrosine	2.00	2.08	2.15	2.22
Minerals:				
Calcium	1.26	1.30	1.34	1.38
Total phosphorus	0.85	0.88	0.91	0.94
Available phosphorus	0.61	0.63	0.65	0.67
Sodium	0.16	0.16	0.17	0.17
Chloride	0.14	0.14	0.15	0.15

Table 48 Recommended dietary concentrations (%) of protein, amino acids and minerals for turkeys during the growing phase (5–8 weeks)

	Dietary energy level (kcal ME/kg)			
	2900	3000	3100	3200
Crude protein	23.2	24.0	24.8	25.6
Amino acids:				
Lysine	1.39	1.44	1.49	1.54
Methionine	0.40	0.41	0.42	0.43
Sulphur amino acids	0.88	0.91	0.94	0.97
Tryptophan	0.21	0.22	0.23	0.24
Threonine	0.80	0.83	0.86	0.89
Glycine + serine	2.22	2.30	2.38	2.46
Leucine	1.35	1.40	1.45	1.50
Isoleucine	0.73	0.75	0.77	0.79
Valine	1.02	1.05	1.08	1.11
Histidine	0.54	0.56	0.58	0.60
Arginine	1.43	1.48	1.53	1.58
Phenylalanine + tyrosine	1.81	1.87	1.93	2.00
Minerals:				
Calcium	1.26	1.30	1.34	1.38
Total phosphorus	0.85	0.88	0.91	0.94
Available phosphorus	0.61	0.63	0.65	0.67
Sodium	0.17	0.17	0.18	0.18
Chloride	0.15	0.15	0.16	0.16

Table 49 Recommended dietary concentrations (%) of protein, amino acids and minerals for turkeys during the growing phase (9–12 weeks)

	Dietary energy level (kcal ME/kg)			
	2900	3000	3100	3200
Crude protein	19.3	20.0	20.7	21.4
Amino acids:				
Lysine	1.11	1.15	1.19	1.23
Methionine	0.33	0.34	0.35	0.36
Sulphur amino acids	0.73	0.75	0.77	0.79
Tryptophan	0.16	0.17	0.18	0.19
Threonine	0.61	0.63	0.65	0.67
Glycine + serine	1.74	1.80	1.86	1.92
Leucine	1.03	1.07	1.11	1.15
Isoleucine	0.55	0.57	0.59	0.61
Valine	0.77	0.80	0.83	0.86
Histidine	0.42	0.43	0.44	0.45
Arginine	1.09	1.13	1.17	1.21
Phenylalanine + tyrosine	1.50	1.55	1.60	1.65
Minerals:				
Calcium	0.97	1.00	1.03	1.06
Total phosphorus	0.72	0.75	0.78	0.81
Available phosphorus	0.48	0.50	0.52	0.54
Sodium	0.15	0.15	0.16	0.16
Chloride	0.14	0.14	0.15	0.15

Table 50 Recommended dietary concentrations (%) of protein, amino acids and minerals for turkeys during the finishing phase (13–16 weeks)

	Dietary energy level (kcal ME/kg)			
	2900	3000	3100	3200
Crude protein	15.5	16.0	16.5	17.0
Amino acids:				
Lysine	0.92	0.95	0.98	1.01
Methionine	0.27	0.28	0.29	0.30
Sulphur amino acids	0.63	0.65	0.67	0.69
Tryptophan	0.14	0.15	0.16	0.17
Threonine	0.48	0.50	0.52	0.54
Glycine + serine	1.55	1.60	1.65	1.70
Leucine	0.83	0.86	0.89	0.92
Isoleucine	0.44	0.46	0.48	0.50
Valine	0.62	0.64	0.66	0.68
Histidine	0.33	0.34	0.35	0.36
Arginine	0.87	0.90	0.93	0.96
Phenylalanine + tyrosine	1.33	1.38	1.43	1.48
Minerals:				
Calcium	0.94	0.97	1.00	1.03
Total phosphorus	0.69	0.72	0.75	0.78
Available phosphorus	0.46	0.48	0.50	0.52
Sodium	0.14	0.15	0.15	0.16
Chloride	0.13	0.14	0.14	0.15

Table 51 Recommended dietary concentrations (%) of protein, amino acids and minerals for turkeys during the finishing phase (17–20 weeks)

	Dietary energy level (kcal ME/kg)			
	2900	3000	3100	3200
Crude protein	13.5	14.0	14.5	15.0
Amino acids:				
Lysine	0.77	0.80	0.83	0.86
Methionine	0.21	0.22	0.23	0.24
Sulphur amino acids	0.48	0.50	0.52	0.54
Tryptophan	0.13	0.13	0.14	0.14
Threonine	0.43	0.44	0.45	0.46
Glycine + serine	1.32	1.37	1.42	1.47
Leucine	0.73	0.75	0.77	0.79
Isoleucine	0.39	0.40	0.41	0.42
Valine	0.54	0.56	0.58	0.60
Histidine	0.29	0.30	0.31	0.32
Arginine	0.76	0.79	0.82	0.85
Phenylalanine + tyrosine	1.24	1.28	1.32	1.36
Minerals:				
Calcium	0.84	0.87	0.90	0.93
Total phosphorus	0.64	0.67	0.70	0.73
Available phosphorus	0.40	0.43	0.45	0.47
Sodium	0.14	0.14	0.15	0.15
Chloride	0.13	0.13	0.14	0.14

Table 52 Recommended dietary concentrations (%) of protein, amino acids and minerals for turkeys during the finishing phase (21–24 weeks)

	Dietary energy level (kcal ME/kg)			
	2900	3000	3100	3200
Crude protein	11.6	12.0	12.4	12.8
Amino acids:				
Lysine	0.58	0.60	0.62	0.64
Methionine	0.19	0.20	0.21	0.22
Sulphur amino acids	0.46	0.48	0.50	0.52
Tryptophan	0.12	0.12	0.13	0.13
Threonine	0.39	0.40	0.41	0.42
Glycine + serine	1.11	1.15	1.19	1.23
Leucine	0.68	0.70	0.72	0.74
Isoleucine	0.35	0.36	0.37	0.38
Valine	0.46	0.48	0.50	0.52
Histidine	0.25	0.26	0.27	0.28
Arginine	0.66	0.68	0.70	0.72
Phenylalanine + tyrosine	1.14	1.18	1.22	1.26
Minerals:				
Calcium	0.74	0.77	0.80	0.83
Total phosphorus	0.62	0.65	0.68	0.71
Available phosphorus	0.39	0.41	0.43	0.45
Sodium	0.14	0.14	0.15	0.15
Chloride	0.13	0.13	0.14	0.14

Table 53 Recommended added dietary levels of trace elements and vitamins for turkeys (mg/kg or iu)

Age: Period:	0–8 weeks Starter/ grower 1	9–16 weeks Grower 2/ finisher 1	16 weeks Finisher 2 + 3	Adult Breeder
Trace elements:				
Iron	40	30	20	40
Copper	4	3	2	3
Zinc	60	40	30	50
Manganese	80	70	40	60
Cobalt	0.2	0.2	0.2	0.2
Selemium	0.15	0.1	0.1	0.15
Iodine	1.0	0.7	0.5	0.8
Vitamins:				
Vitamin A (iu)	10 000	8 000	8 000	10 000
Vitamin D_3 (iu)	1 500	1 200	1 200	1 500
Vitamin E	20	15	10	15
Vitamin K_3	4	3	2	4
Thiamine	2	1	–	1
Riboflavin	6	4	4	6
Pantothenic acid	10	5	5	10
Niacin	60	40	40	40
Pyridoxine	2	–	–	2
Biotin	0.3	0.05	–	0.15
Folic acid	1.2	0.7	0.5	1.0
Vitamin B_{12}	0.015	0.010	0.010	0.010
Choline chloride	800	800	500	600

Table 54 Recommended dietary concentrations (%) of protein, amino acids and minerals for turkey breeders during the rearing phase

	Period		
	0–4 weeks	5–14 weeks	15–25 weeks
Dietary energy level (kcal ME/kg):	3000	2900	2900
Crude protein	24.0	18.0	13.5
Amino acids:			
Lysine	1.44	0.95	0.75
Methionine	0.41	0.31	0.20
Sulphur amino acids	0.91	0.70	0.47
Tryptophan	0.22	0.15	0.13
Threonine	0.83	0.60	0.42
Minerals:			
Calcium	1.30	1.00	0.84
Total phosphorus	0.88	0.72	0.64
Available phosphorus	0.63	0.48	0.40
Sodium	0.17	0.15	0.14
Chloride	0.15	0.14	0.13

Energy

As it does not, in contrast to the chicken, have a tendency to lay down fat, the turkey may be fed diets of relatively high energy concentration: in addition as it adapts very well to less concentrated diets, the range of dietary energy levels may be relatively large (2500–3000 kcal ME/kg). The use of quantitative feed restriction is not justified since the change in energy requirements is automatically associated with an alteration in intake.

Protein, amino acids and minerals

Daily requirements of the breeder turkey are collected together in *Table 55*. The values given correspond to their period of maximum production. Taking into account these requirements and the intakes usually observed, it is recommended that a feed having the characteristics indicated in *Table 56* be used. Recommended dietary levels of trace elements and vitamins are presented in *Table 53*, with respect to those relative to the growing bird.

Table 55 Daily allowances of energy, crude protein and amino acids for turkey breeders

	Type		
	Farm-type	*Medium*	*Heavy*
Energy (kcal ME)	460	620	690
Crude protein (g)		27.0	
Amino acids (g):			
Lysine		1.28	
Methionine		0.78	
Sulphur amino acids		1.05	
Tryptophan		0.28	
Threonine		0.89	
Minerals:			
Calcium*		4.5/6.0	
Total phosphorus		1.25	
Available phosphorus		0.75	
Sodium		0.30	
Chloride		0.25	

*The initial figure refers to lay, whereas the second is to ensure optimum hatchability.

Note Males are usually reared separately from females; information currently available on their nutrition is insufficient to allow precise recommendations.

Table 56 Recommended dietary concentrations (%) of protein, amino acids and minerals for turkey breeders*

	Dietary energy level (kcal ME/kg)	
	2800	*3000*
Crude protein	13.0	14.0
Amino acids:		
Lysine	0.60	0.64
Methionine	0.35	0.37
Sulphur amino acids	0.48	0.51
Tryptophan	0.13	0.14
Threonine	0.40	0.43
Minerals:		
Calcium**	2.30/3.20	2.45/3.40
Total phosphorus	0.63	0.65
Available phosphorus	0.38	0.40
Sodium	0.15	0.16
Chloride	0.13	0.14
Expected daily feed intake (g) at 18°C	160–230†	150–210†

*See *Table 53* (adult) for trace elements and vitamins.
**See footnote to *Table 55*.
†Based on genotype and length of lay.

Chapter 12
Nutrition of guinea-fowl

Meat guinea-fowl

Guinea-fowl are reared for between 11 and 13 weeks. In order to match dietary levels to nutrient requirements, three feeds must be used, corresponding respectively to the three periods 0 to 4 weeks (starter), 5 to 8 weeks (grower) and 9 to 11 or 13 weeks (finisher).

As a guideline, *Table 57* presents liveweight gain, feed intake and feed conversion ratio measured at the end of these periods, and *Table 58* gives weekly feed and water consumption.

Body development is the same in both sexes up until 12 weeks of age. Beyond this, liveweight gain in females is some 20% higher, which is associated with sexual maturity (influenced by lighting patterns) and corresponds only to fat depots and development of genitalia.

Table 57 Performance characteristics of guinea-fowl (for period indicated)

Period (week)	*Liveweight gain* (g)	*Feed intake** (g)	*Feed conversion ratio**
0–4	380	670	1.76
5–8	590	1690	2.86
9–11	400	1735	4.34
12	110	630	5.73
13	100	635	6.35
0–11	1370	4095	2.99
0–12	1480	4725	3.19
0–13	1580	5360	3.39

*Dietary energy level of 3000 kcal ME/kg and an ambient temperature of 20°C.

Presentation and energy level of the diet

It is recommended that initially crumbs be used followed by, at two weeks of age, pellets (2.5 mm diameter). Invariably, and in contrast to the broiler, guinea-fowl may be perfectly adequately fed meal up to eight weeks of age and neither growth rate nor feed conversion ratio (if wastage is avoided) are affected.

Table 58 Weekly feed and water intakes of the growing guinea-fowl

Week	Feed (g)*	Water (g)
1	55	75
2	130	180
3	210	290
4	275	380
5	340	460
6	395	530
7	460	620
8	495	670
9	540	730
10	585	780
11	610	810
12	630	835
13	635	845
14	640	850

*Dietary energy level of 3000 kcal ME/kg at an ambient temperature of 20°C.

Growth rate of guinea-fowl may be slightly altered if the dietary energy value is changed; beyond 2900 kcal ME/kg an increase of 100 kcal ME/kg only improves growth rate by 1% at 12 weeks and by 1.1% at 13 weeks. Again, the gain in weight is associated with fat deposition. For the production of lean guinea-fowl that are nevertheless old (13–14 weeks) males only must be kept and fed a diet of relatively low energy level (2700–2900 kcal ME/kg). In all cases, high energy levels are costly and are not compensated for by the improvement in feed conversion ratio; therefore the level which corresponds to the lowest unit energy cost must be used.

Protein and amino acids

Protein, lysine and sulphur amino acid requirements presented in *Table 59* correspond to the liveweight gain at the end of each period. They are expressed in absolute terms as a total requirement for each period of growth. The values relating to protein levels may seem low; they presuppose that synthetic lysine supplementation is possible and would only be effective if such supplementation were to be economic.

Recommended levels for starter, grower and finisher periods (*Tables 60 to 62*) take these requirements into account and in the main allow the levels of performance outlined in *Table 57* to be achieved. They may nevertheless be increased if, at a similar energy level, a significant reduction in food intake is anticipated as a consequence of a higher ambient temperature (feed conversion ratio from 0 to 12 weeks better than 3.05 at 3000 kcal ME/kg or better than 2.85 at 3200 kcal ME/kg). If accidentally (deficiency, sickness) growth rate is retarded, the use of the diet corresponding to the younger age (starter or grower) may be prolonged up until the point where performance has caught up again.

Finally it may be noted that the balance between sulphur amino acids and lysine must be higher with the guinea-fowl than in the broiler.

Table 59 Allowances of crude protein and amino acids for the growing guinea-fowl (g/bird per period)

Period (week)	Crude protein	Lysine	Sulphur amino acids
0–4	165	9.0	6.4
5–8	295	14.8	13.7
9–11	225	11.3	10.9
12	70	3.0	3.0
13	70	3.0	3.0
0–11	685	35.1	31
0–12	755	38.1	34
0–13	825	41.1	37

Table 60 Recommended dietary concentrations (%) of crude protein, amino acids and minerals for the guinea-fowl during the starter phase (0–4 weeks)

	Dietary energy level (kcal ME/kg)			
	2900	3000	3100	3200
Crude protein	23.2	24.0	24.8	25.5
Amino acids:				
Lysine	1.26	1.30	1.34	1.38
Methionine	0.52	0.53	0.54	0.55
Sulphur amino acids	0.92	0.95	0.98	1.00
Tryptophan	0.23	0.24	0.24	0.25
Threonine	0.82	0.85	0.88	0.90
Minerals:				
Calcium	1.00	1.03	1.06	1.10
Total phosphorus	0.64	0.65	0.66	0.67
Available phosphorus	0.39	0.40	0.41	0.42
Sodium	0.16	0.16	0.17	0.17
Chloride	0.14	0.14	0.15	0.15

Minerals and vitamins

Guinea-fowl raised on litter have no mineral requirements higher than those of the broiler. The usual recommendations are therefore presented in *Tables 60* to *62* with respect to phosphorus, calcium, sodium and chloride.

Additions of trace elements indicated in *Table 63* are associated with a margin of security to account for raw material variability. This table also includes recommended additions of vitamins. A high level of vitamin E and niacin during the starter phase is justified to avoid all risks of perosis.

Breeders during rearing

Adipose tissue in the female guinea-fowl tends to develop from ten weeks of age and must be restricted during this period of rearing. This practice is not associated

Table 61 Recommended dietary concentrations (%) of crude protein, amino acids and minerals for the guinea-fowl during the growing phase (5–8 weeks)

	Dietary energy level (kcal ME/kg)			
	2800	*2900*	*3000*	*3100*
Crude protein	17.3	17.9	18.5	19.1
Amino acids:				
Lysine	0.89	0.92	0.95	0.98
Methionine	0.38	0.39	0.40	0.42
Sulphur amino acids	0.78	0.80	0.83	0.86
Tryptophan	0.20	0.21	0.22	0.23
Threonine	0.68	0.70	0.72	0.74
Glycine + serine	1.52	1.57	1.63	1.68
Leucine	1.56	1.62	1.67	1.72
Isoleucine	0.78	0.81	0.84	0.87
Valine	0.86	0.89	0.92	0.95
Histidine	0.43	0.45	0.46	0.48
Arginine	1.11	1.15	1.19	1.23
Phenylalanine + tyrosine	1.51	1.56	1.62	1.67
Minerals:				
Calcium	0.87	0.90	0.93	0.96
Total phosphorus	0.59	0.60	0.61	0.62
Available phosphorus	0.34	0.35	0.36	0.37
Sodium	0.16	0.16	0.17	0.17
Chloride	0.14	0.14	0.15	0.15

Table 62 Recommended dietary concentrations (%) of crude protein, amino acids and minerals for the guinea-fowl during the finishing phase (9–12 weeks)

	Dietary energy level (kcal ME/kg)			
	2800	*2900*	*3000*	*3100*
Crude protein	12.8	13.3	13.7	14.1
Amino acids:				
Lysine	0.60	0.62	0.64	0.66
Methionine	0.30	0.31	0.32	0.33
Sulphur amino acids	0.60	0.62	0.64	0.66
Tryptophan	0.11	0.12	0.12	0.13
Threonine	0.48	0.50	0.51	0.53
Glycine + serine	1.13	1.17	1.21	1.25
Leucine	1.38	1.43	1.48	1.53
Isoleucine	0.55	0.57	0.59	0.61
Valine	0.64	0.67	0.69	0.72
Histidine	0.33	0.34	0.35	0.36
Arginine	0.74	0.77	0.79	0.82
Phenylalanine + tyrosine	1.19	1.23	1.27	1.31
Minerals:				
Calcium	0.77	0.80	0.83	0.86
Total phosphorus	0.56	0.57	0.58	0.59
Available phosphorus	0.31	0.32	0.33	0.34
Sodium	0.16	0.16	0.17	0.17
Chloride	0.14	0.14	0.15	0.15

Table 63 Recommended added dietary levels of trace elements and vitamins for the guinea-fowl (mg/kg or iu)

	Starter	Grower/finisher
Trace elements (ppm):		
Iron	25	15
Copper	3	2
Zinc	40	25
Manganese	70	50
Cobalt	0.15	–
Selenium	0.15	–
Iodine	1	1
Vitamins:		
Vitamin A (iu/kg)	12 000.00	10 000
Vitamin D_3 (iu/kg)	2 000	1 000
Vitamin E	25	12
Vitamin K_3	3	2
Riboflavin	5	5
Pantothenic acid	8	8
Niacin	30	15
Pyridoxine	1	–
Biotin	0.2	–
Folic acid	0.2	–
Vitamin B_{12}	0.01	0.01
Choline chloride	500	250

Table 64 Recommended dietary concentrations (%) of crude protein, amino acids and minerals for the breeder guinea-fowl during the rearing phase

	Period		
	Starter 0–4 weeks	Grower 5–12 weeks	Rearer 13–22 weeks
Dietary energy levels (kcal ME/kg)	*2900*	*2800*	*2800*
Crude protein	20.0	14.0	12.0
Amino acids:			
Lysine	1.20	0.55	0.48
Methionine	0.40	0.28	0.22
Sulphur amino acids	0.85	0.60	0.50
Tryptophan	0.25	0.14	0.12
Minerals:			
Calcium	0.85	0.80	0.50
Total phosphorus	0.65	0.60	0.50
Available phosphorus	0.40	0.35	0.25
Sodium	0.17	0.17	0.17
Chloride	0.15	0.15	0.15

with any adverse effect on subsequent lay and, on the contrary, reduces mortality at the point of lay; restriction results only in a delay in the attainment of sexual maturity which is easily compensated for by the lighting programme.

The principles governing feed restriction in pullets may be adopted for the guinea-fowl. In practice the growth rate may be already modified during the first 10–12 weeks by reducing dietary protein levels. Characteristics of the diets to be used are presented in *Table 64*. Beyond this stage (10–12 weeks) quantitative restriction is more economical. It will be noted that daily feed levels are featured in *Table 65*.

Table 65 Feed allowances for the breeder guinea-fowl during the rearing phase*

Age (weeks)	Type of diet	Feed (g/bird per day)	Cumulative feed consumption (g/bird)
0–4	Starter	ad libitum	500
5–12	Grower	ad libitum	3000
13–15	Rearer	55	4155
16–17	–	63	5040
18–22	–	72	7560
23	–	80	8120
From 24 until point of lay		85	
From 25% rate of lay		ad libitum**	

*Dietary energy level of 2800 kcal ME/kg at a mean ambient temperature of 18°C.
**It may be better to maintain a lower level of intake during lay.

Breeders

The best performances in lay are achieved with dietary energy levels equal to or higher than 2800 kcal ME/kg.

The daily nutrient requirements during the month when rate of lay is highest, are given in *Table 66*. Taking these values into account, and normal rearing conditions, dietary characteristics featured in *Table 67* may be recommended. They will promote at the same time both maximum rate of lay and hatchability.

Table 66 Daily energy and nutrient allowances of breeder guinea-fowl

ME (kcal)	310 (dependent upon ambient temperature)
Crude protein (g)	13.0
Amino acids (g):	
Lysine	0.58
Methionine	0.30
Sulphur amino acids	0.53
Tryptophan	0.14
Threonine	0.41
Glycine + serine	1.20
Leucine	0.75
Isoleucine	0.62
Valine	0.51
Histidine	0.17
Arginine	0.54
Phenylalanine + tyrosine	0.97
Minerals (g):	
Calcium	3.8
Total phosphorus	0.68
Available phosphorus	0.44
Sodium	0.14
Chloride	0.13
Linoleic acid (g)	0.80

Table 67 Recommended dietary concentrations (%) of crude protein, amino acids and minerals for the breeder guinea-fowl*

	Dietary energy level (kcal ME/kg)		
	2800	*2900*	*3000*
Crude protein	13.5	14.0	14.5
Amino acids:			
Lysine	0.77	0.80	0.83
Methionine	0.30	0.31	0.32
Sulphur amino acids	0.54	0.56	0.57
Tryptophan	0.14	0.15	0.15
Threonine	0.36	0.37	0.38
Minerals:			
Calcium	3.70	3.85	4.00
Total phosphorus	0.67	0.68	0.70
Available phosphorus	0.42	0.44	0.45
Sodium	0.14	0.15	0.15
Chloride	0.13	0.14	0.14
Based on an expected daily feed intake (g) at an ambient temperature of 20°C:			
30–50 weeks of age	105	102	100
51–70 weeks of age	100	97	95

*In the absence of experimental data, concentrations of trace elements and vitamins may follow those for the breeder turkey (*Table 53*).

Chapter 13
Nutrition of ducks

Types of meat duck raised in France

Ducks raised for meat production are from two different species (*Table 68*): the common duck, *Anas platirynchos*, which comprises diverse types that are also occasionally used in crossing (Peking, Kaki, Rouen, Wild Cross for example) and the Barbary or Muscovy duck, *Cairina moschata*.

When compared with broilers, the two species have a rapid liveweight gain but the development of pectoral muscle tissue (breast meat) does not take place until relatively later and at an age when fattening is becoming important.

Sexual dimorphism is not very important in the common duck. Slaughter may be between seven and eight weeks for both sexes, taking into account the late development of muscle tissue and the slowing down of liveweight gain after seven

Table 68 Performance characteristics of ducks*

Type		\multicolumn{7}{c}{Age (weeks)}						
		2	4	7	8	9	10	11	12
Peking	Liveweight (kg)	0.50	1.32	2.30	2.44	2.54			
	Cumulative feed intake (kg)	0.76	2.85	6.77	8.02	9.17			
	Cumulative feed conversion ratio	1.70	2.25	3.00	3.35	3.68			
Peking Cross	Liveweight (kg)	0.30	0.98	2.00	2.10				
	Cumulative feed intake (kg)	0.39	2.05	6.00	6.70				
	Cumulative feed conversion ratio	1.55	2.20	3.10	3.30				
Wild Cross**	Liveweight (kg)	0.27			1.40				
	Cumulative feed intake (kg)	0.35			5.40				
	Cumulative feed conversion ratio	1.55			4.00				
Barbary male	Liveweight (kg)	0.32	1.05	2.55	2.90	3.22	3.55	3.65	3.75
	Cumulative feed intake (kg)	0.30	1.78	5.52	6.90	8.24	9.54	10.84	12.14
	Cumulative feed conversion ratio	1.20	1.78	2.21	2.42	2.60	2.73	3.00	3.28
Barbary female	Liveweight (kg)	0.28	0.95	1.74	2.00	2.08	2.15	2.20	
	Cumulative feed intake (kg)	0.30	1.45	4.14	5.04	5.84	6.66	7.43	
	Cumulative feed conversion ratio	1.30	1.61	2.45	2.58	2.88	3.17	3.46	

*At a dietary energy level of 2900 kcal ME/kg and an ambient temperature of 18°C
**Cross between wild drake and Peking duck.

weeks. On the other hand, sexual dimorphism is very evident in the Barbary duck (*Table 68*). Fat depots are smaller and muscle mass more important towards the end of growth than with the common duck. After 10 weeks in females and 11 weeks in males, development slows down considerably, but the major component of liveweight gain at this time (55%) corresponds to growth of the pectoral muscles. This development occurs therefore at a time when the curve of liveweight gain reaches inflection and when partial feed conversion ratio (calculated over a short time period) worsens dramatically and may approach values higher than eight. As a consequence, a relatively early or late slaughter age must be selected according to whether a good price or a fleshy carcass is required. It is recommended that females should never be slaughtered before 70 days and males never before 77 days.

Nutrition of roasting ducks

Despite important morphological differences, the two species of duck have reasonably similar nutrient requirements which will not be distinguished subsequently. *Table 69* includes feed consumption of each species. Water consumption is not precisely known because of continuous wastage.

Table 69 Weekly feed intake (g) of ducks*

Week	Peking	Barbary male	Barbary female
1	235	60	60
2	525	240	210
3	910	560	430
4	1175	920	720
5	1300	1150	875
6	1320	1250	900
7	1300	1340	920
8	1250	1380	900
9	1150	1340	800
10	1100	1300	800
11		1300	770
12		1300	

*Dietary energy level of 2900 kcal ME/kg.

Dietary energy concentration

The duckling is well able to adjust its feed intake to meet its energy requirements. With an increase in dietary energy level from 2400 to 3200 kcal ME/kg, slaughter weight is unchanged and carcass fat levels are barely higher. Therefore interest is centred on using those diets which correspond to the lowest unit energy cost. In practice, levels between 2800 and 3000 kcal ME/kg are adopted and raw material energy values as determined with the broiler may be used.

Protein and amino acids

Ducklings have lower protein requirements than broilers. To ensure a maximum growth rate during the first week of life they are certainly high but, as a consequence of a remarkable compensatory growth rate, ducks may reach the same final liveweight with a slower growth rate during the starter period. The use of high protein levels is therefore not justified; they will not improve feed conversion ratio and will only marginally reduce carcass fat levels. It is essential to ensure that sulphur amino acid requirements are met such that maximum efficiency of feed utilization is achieved.

It is probable that the requirements of the two sexes of the common duck are similar. On the other hand the male Barbary duck has a longer growth period and, as a consequence, has a higher protein requirement until a later age than that of females.

Protein and amino acid levels designed to promote the best performance are indicated in *Tables 70* to *74*. The values for protein assume that there is the possibility of supplementation with synthetic lysine. In practice these norms, which may be a little low, are usually higher because of the need to meet lysine requirements. They would only be effective if the use of synthetic lysine were economic.

Two types of diet (starter, finisher) are suggested for the common duck and three for the Barbary which needs an intermediary diet (grower) for a variable length of time depending upon sex. If the two sexes are reared together, the finisher diet designed for the males must be used.

Table 70 Recommended dietary energy levels and concentrations (%) of crude protein, amino acids and minerals for starter ducks (0–14 days for common ducks: Peking and Wild Cross; and 0–21 days for the Barbary duck)

	Dietary energy level (kcal ME/kg)		
	2600	*2800*	*3000*
Crude protein	16.5	17.7	19.0
Amino acids:			
Lysine	0.83	0.90	0.96
Methionine	0.36	0.38	0.41
Sulphur amino acids	0.70	0.75	0.80
Tryptophan	0.18	0.19	0.20
Threonine	0.60	0.65	0.69
Glycine + serine	1.44	1.54	1.64
Leucine	1.58	1.69	1.80
Isoleucine	0.75	0.80	0.85
Valine	0.81	0.87	0.93
Histidine	0.41	0.44	0.47
Arginine	0.96	1.03	1.10
Phenylalanine + tyrosine	1.47	1.57	1.67
Minerals:			
Calcium	0.80	0.85	0.90
Total phosphorus	0.61	0.63	0.65
Available phosphorus	0.36	0.38	0.40
Sodium	0.14	0.15	0.16
Chloride	0.12	0.13	0.14

Table 71 Recommended dietary energy levels and concentrations (%) of crude protein and amino acids for the common duck during the finishing phase (15 days to slaughter)

	Dietary energy level (kcal ME/kg)		
	2600	*2800*	*3000*
Crude protein	12.6	13.6	14.5
Amino acids:			
Lysine	0.63	0.68	0.72
Methionine	0.27	0.29	0.31
Sulphur amino acids	0.54	0.58	0.62

Table 72 Recommended dietary energy levels and concentrations (%) of crude protein, amino acids and minerals for the Barbary duck during the growing phase (22–41 days)

	Dietary energy level (kcal ME/kg)		
	2600	*2800*	*3000*
Crude protein	13.9	14.9	16.0
Amino acids:			
Lysine	0.66	0.71	0.76
Methionine	0.29	0.31	0.33
Sulphur amino acids	0.57	0.61	0.65
Tryptophan	0.14	0.15	0.16
Threonine	0.48	0.51	0.55
Glycine + serine	1.13	1.22	1.30
Leucine	1.24	1.34	1.43
Isoleucine	0.58	0.62	0.67
Valine	0.64	0.69	0.74
Histidine	0.32	0.35	0.37
Arginine	0.80	0.86	0.92
Phenylalanine + tyrosine	1.15	1.23	1.32
Minerals:			
Calcium	0.70	0.75	0.80
Total phosphorus	0.55	0.58	0.60
Available phosphorus	0.30	0.33	0.35
Sodium	0.14	0.15	0.16
Chloride	0.12	0.13	0.14

Ducklings are capable of withstanding temperature changes after three weeks of age but in doing so modify their feed consumption. This is particularly the case with temperatures below 10°C which raise energy expenditure and accordingly increase appetite. Feed intake therefore is very variable according to rearing conditions (summer or winter, within buildings or semi-extensive) whereas protein requirements are dependent above all on liveweight gain. Recommended dietary levels

Table 73 Recommended dietary energy levels and concentrations (%) of crude protein, amino acids and minerals for the female Barbary duck during the finishing phase (42 days until slaughter)

	Dietary energy level (kcal ME/kg)		
	2600	2800	3000
Crude protein	11.3	12.2	13.0
Amino acids:			
Lysine	0.50	0.54	0.58
Methionine	0.21	0.23	0.24
Sulphur amino acids	0.43	0.46	0.50
Tryptophan	0.10	0.11	0.12
Threonine	0.35	0.38	0.41
Glycine + serine	0.86	0.93	1.00
Leucine	0.97	1.05	1.13
Isoleucine	0.44	0.47	0.51
Valine	0.49	0.53	0.57
Histidine	0.25	0.27	0.29
Arginine	0.60	0.65	0.70
Phenylalanine + tyrosine	0.89	0.96	1.03
Minerals:			
Calcium	0.60	0.65	0.70
Total phosphorus	0.47	0.49	0.51
Available phosphorus	0.22	0.24	0.26
Sodium	0.14	0.15	0.16
Chloride	0.12	0.13	0.14

Table 74 Recommended dietary energy levels and concentrations (%) of crude protein, amino acids and minerals for male or as-hatched Barbary ducks during the finishing phase (42 days until slaughter)

	Dietary energy level (kcal ME/kg)		
	2600	2800	3000
Crude protein	12.1	13.0	14.0
Amino acids:			
Lysine	0.60	0.65	0.70
Methionine	0.22	0.24	0.26
Sulphur amino acids	0.47	0.50	0.54
Tryptophan	0.12	0.13	0.14
Threonine	0.22	0.24	0.26
Glycine + serine	1.04	1.12	1.21
Leucine	1.17	1.26	1.36
Isoleucine	0.53	0.57	0.61
Valine	0.59	0.64	0.69
Histidine	0.30	0.33	0.35
Arginine	0.72	0.78	0.84
Phenylalanine + tyrosine	1.07	1.15	1.24
Minerals:			
Calcium	0.60	0.65	0.70
Total phosphorus	0.47	0.49	0.51
Available phosphorus	0.22	0.24	0.26
Sodium	0.14	0.15	0.16
Chloride	0.12	0.13	0.14

have been calculated on the basis of the intake and performance data presented in *Tables 68* and *69*; they must be modified if intake is altered by various factors. It is for this reason that the considerable reduction in appetite observed beyond an ambient temperature of 22°C may justify the use of protein-rich diets: an elevation of 2% in amino acid levels per degree increase in temperature (beyond 20°C) is usually sufficient.

Minerals and vitamins

Recommendations concerning dietary levels of calcium, phosphorus, sodium and chloride, sufficient to meet requirements but no more, are given in *Tables 70* to *74*. On the other hand, recommended added amounts of trace elements and vitamins presented in *Table 75* do contain a margin of security.

Table 75 Recommended added dietary levels of trace elements and vitamins for the duck (mg/kg or iu)

	Starter	*Grower (Barbary duck) and finisher (common duck)*	*Finisher (Barbary duck)*
Trace elements:			
Iron	40	30	20
Copper	5	4	3
Zinc	40	30	20
Manganese	70	60	60
Cobalt	0.2	0.2	0.2
Selenium	0.1	0.1	0.1
Iodine	1.0	0.7	0.5
Vitamins:			
Vitamin A (iu)	8000	8000	4000
Vitamin D_3 (iu)	1000	1000	500
Vitamin E	20	15	–
Vitamin K_3	4	4	–
Thiamine	1	–	–
Riboflavin	4	4	2
Pantothenic acid	5	5	–
Niacin	25	25	–
Pyridoxine	2	–	–
Biotin	0.1	–	–
Folic acid	0.2	–	–
Vitamin B_{12}	0.03	0.01	–
Choline chloride	300	300	–

Presentation of diets

The most commonly used forms are crumbs for the starter diets and pellets 3–5 mm in diameter for grower and finisher feeds.

Ducklings may, however, successfully consume meal diets if these are fed from hatching but their use is associated with a reduction in liveweight of 5% and has to be accompanied by specific precautions if wastage is to be avoided.

Nutrition of breeders during rearing

Between six (common duck) or ten weeks of age (Barbary duck) and point of lay, ducks must be fed on a restricted basis if it is wished to delay the onset of sexual maturity: the use of lighting regimes, even if properly designed, is in fact insufficient to control this parameter in web-footed birds. The level of restriction selected is usually a function of liveweight gain and is controlled fortnightly by weighing a sample of birds. It is in the region of 100–140 g/day (depending upon type of bird) for females, and 150–160 g/day for Barbary males.

Nutrition of breeder ducks

Daily requirements for breeder ducks are presented in *Table 76*; *Table 77* contains subsequent dietary levels taking into account levels of intake usually recorded.

Table 76 Daily energy and nutrient allowances for the breeder duck during lay*

	Barbary (2.5 kg liveweight)	Peking (2.3 kg liveweight)
ME (kcal)	425	Variable according to genotype
Crude protein (g)	18.0	17.0
Amino acids (g):		
Lysine	0.88	0.85
Methionine	0.52	0.44
Sulphur amino acids	0.89	0.75
Tryptophan	0.21	0.20
Minerals (g):		
Calcium	4.20	4.20
Total phosphorus	0.90	0.90
Available phosphorus	0.60	0.60
Sodium	0.20	0.20
Chloride	0.20	0.20

*Not including males.

Energy levels

Daily energy requirements of common ducks obviously vary as a function of their body size. The requirements of Barbary ducks are slightly higher than 400 kcal ME at an ambient temperature of between 15 and 20°C.

As with young, adult ducks are able to regulate their energy intake perfectly if fed diets varying between 2400 and 3000 kcal ME/kg. However, dietary energy levels for Barbary breeders ought to be higher than 2800 kcal ME/kg in order to maximize fertility and hatchability.

Table 77 Recommended dietary energy levels and concentrations (%) of crude protein, amino acids and minerals for breeder ducks during lay*

	Barbary		Peking	
Dietary energy level (kcal ME/kg):	2600	2800	2600	2800
Crude protein	12.00	13.00	13.00	14.00
Amino acids:				
Lysine	0.62	0.66	0.66	0.71
Methionine	0.33	0.35	0.33	0.35
Sulphur amino acids	0.56	0.60	0.56	0.60
Tryptophan	0.15	0.16	0.16	0.17
Threonine	0.43	0.46	0.46	0.49
Minerals:				
Calcium	2.50	2.70	2.50	2.70
Total phosphorus	0.60	0.62	0.60	0.62
Available phosphorus	0.37	0.40	0.37	0.40
Sodium	0.14	0.15	0.14	0.15
Chloride	0.13	0.14	0.13	0.14
Daily food intake (g) at 18°C	165	155		

*In the absence of experimental data, concentrations of trace elements and vitamins may follow those for the breeder turkey (*Table 53*).

Protein and amino acids

Daily requirements of common ducks are variable principally because of the diversity of birds all with this same name; the values given correspond to maximum performance levels. Similarly protein requirements for Barbary breeders presented in *Table 76* correspond to that period when egg production is at its highest.

Chapter 14

Nutrition of geese

Geese reared in France are used for force-feeding, the production of meat for roasting and finally for breeding.

Geese grow extremely rapidly during the first eight weeks of life. Beyond this age, liveweight gain is considerably slower and after 12 weeks is practically zero. Sexual dimorphism is only moderate but detectable from four weeks of age. Growth rate and mean feed intake of geese are presented in *Table 78*; there is little difference between the principal types (Landaise, Rhine).

Table 78 Performance characteristics and grass consumption of geese*

	Age (weeks)						
	0	*2*	*4*	*6*	*8*	*10*	*12*
Males:							
Liveweight (g)	82	860	2 300	3 900	4 850	5 020	5 200
Cumulative feed intake (g)			850	3 500	7 300	11 300	14 800	18 300
Cumulative grass consumption (g)			500	3 800	8 900	14 800	20 400	26 000
Females:							
Liveweight (g)	80	830	2 050	3 350	4 200	4 370	4 530
Cumulative feed intake (g)		830	3 100	6 400	9 600	12 550	15 500
Cumulative grass consumption (g)			500	3 800	8 900	14 800	20 400	26 000

*Dietary energy level of 2800 kcal ME/kg.

Roasting geese

Geese for roasting are slaughtered between 9 and 11 weeks of age. In all cases a rapid growth rate and low body fat content are required. The dietary energy level has no effect on growth rate if it is between 2500 and 3000 kcal ME/kg; in practice levels below 2900 kcal ME/kg are used. It is advisable to exclude fat as a dietary raw material in order to avoid excessive carcass fat levels.

Amino acid and mineral requirements are presented in *Table 79*. It is possible to use several nutritional programmes bearing in mind the capacity of geese for

Table 79 Recommended dietary energy levels and concentrations (%) of crude protein, amino acids and minerals for geese*

	\multicolumn{6}{c}{Age (weeks)}					
	0–3		4–6		7–12	
Dietary energy level (kcal ME/kg):	2600	2800	2700	2900	2700	2900
Crude protein	15.8	17.0	11.6	12.5	10.2	11.0
Amino acids:						
Lysine	0.89	0.95	0.56	0.60	0.47	0.50
Methionine	0.40	0.42	0.29	0.31	0.25	0.27
Sulphur amino acids	0.79	0.85	0.56	0.60	0.48	0.52
Tryptophan	0.17	0.18	0.13	0.14	0.12	0.13
Threonine	0.58	0.62	0.46	0.49	0.43	0.46
Minerals:						
Calcium	0.75	0.80	0.75	0.80	0.65	0.70
Total phosphorus	0.67	0.70	0.62	0.65	0.57	0.60
Available phosphorus	0.42	0.45	0.37	0.40	0.32	0.35
Sodium	0.14	0.15	0.14	0.15	0.14	0.15
Chloride	0.13	0.14	0.13	0.14	0.13	0.14

*In the absence of experimental data, concentrations of trace elements and vitamins may follow those for the duck (*Table 75*).

compensatory growth rate beyond four weeks of age. In order to adjust dietary levels to meet requirements it is recommended that several successive diets be used from hatching to three weeks, from four to six weeks, and from seven weeks to slaughter.

Problems of cannibalism may be reduced if fresh grass is offered with compound diets or if geese are allowed access to grass runs.

Force-fed and rearing breeder geese

It is not possible with current knowledge to indicate whether nutrient requirements for these two types are any different from those for roasting geese. It may be noted that it is beneficial to restrict the amount of feed offered from eight weeks of age to 180 g/day if the dietary energy level is 2750 kcal ME/kg: the use of any other dietary energy level requires an appropriate recalculation of the amount to be fed. The amount of grass offered should be limited to 400 g/day when the birds are not free range.

It is equally possible to restrict birds from hatching, at a level of 90% of the observed *ad libitum* intake. At eight weeks of age the programme described above is adopted.

Breeder geese

During the season of lay, geese need a compound pelleted feed and fresh grass simultaneously for maximum production. The amount of grass given if geese are housed in huts may be restricted to 700–800 g/day without problems. The complete removal of grass is possible if the compound feed is fortified with protein, potassium and certain vitamins but this is always associated with the loss of around two goslings per goose. The amount of compound feed consumed is very variable, dependent upon the stage of lay, with an average of 240 g/day at a dietary energy concentration of 2500 kcal ME/kg.

The characteristics of the compound feed to be used are given in *Table 80*.

Table 80 **Recommended dietary energy levels and concentrations (%) of crude protein, amino acids and minerals for breeder geese**

	Dietary energy level (kcal ME/kg)	
	2200	*2500*
Crude protein	13.0	14.8
Amino acids:		
Lysine	0.58	0.66
Methionine	0.23	0.26
Sulphur amino acids	0.42	0.47
Tryptophan	0.13	0.15
Threonine	0.40	0.45
Minerals:		
Calcium	2.60	3.00
Total phosphorus	0.56	0.60
Available phosphorus	0.32	0.36
Sodium	0.12	0.14
Chloride	0.12	0.14
Daily feed intake (g):		
Point of lay	170	150
End of lay	350	300

Chapter 15
Nutrition of Japanese quail

Growing quail

Mean growth rate and feed intake for growing quail are presented in *Table 81*. Growth rate is rapid until five weeks of age; thereafter it slows down and adult weight is achieved at 50 days. Slaughter is between 40 and 45 days. In contrast to the situation found with the majority of domestic avian species, sexual dimorphism in quail favours the female.

Table 81 Performance characteristics of quail

	\multicolumn{6}{c}{Age (days)}					
	\multicolumn{2}{c}{21}	\multicolumn{2}{c}{40}	\multicolumn{2}{c}{45}
	Male	Female	Male	Female	Male	Female
Liveweight (g)	90	95	125	155	135	165
Feed intake for period* (g)	285	300	210	310	75	80
Cumulative feed consumption (g)	285	300	495	610	570	690
Cumulative feed conversion ratio	–	–	4.12	4.07	4.38	4.31

*0–20 days; 21–40 days; 41–45 days.

Similar growth rates are achieved with quail fed diets ranging from 2600 to 3200 kcal ME/kg. Nevertheless, the need to produce carcasses with adequate fat cover frequently requires the use of diets with a high energy concentration.

Protein and amino acid requirements expressed as a percentage of the diet are very high at the beginning of life and fall rapidly thereafter. There are many possibilities concerning the succession of diets to be used; in practice a starter diet may be fed until 21 days of age followed by a finisher until slaughter.

Among minerals and vitamins, the only elements for which the quail may have a non-typical requirement are choline and zinc.

Recommended characteristics of the two diets are given in *Tables 82* and *83*.

Table 82 Recommended dietary energy levels and concentrations (%) of crude protein, amino acids and minerals for the quail during the starter phase (0–21 days)*

	Dietary energy level (kcal ME/kg)		
	2800	*3000*	*3200*
Crude protein	23.00	24.60	26.30
Amino acids:			
Lysine	1.30	1.39	1.48
Methionine	0.39	0.42	0.45
Sulphur amino acids	0.85	0.91	0.97
Tryptophan	0.20	0.21	0.22
Threonine	0.75	0.80	0.85
Glycine + serine	2.10	2.25	2.40
Leucine	1.28	1.34	1.46
Isoleucine	0.67	0.72	0.77
Valine	0.94	1.00	1.06
Histidine	0.51	0.55	0.59
Arginine	1.32	1.41	1.50
Phenylalanine + tyrosine	1.75	1.87	1.99
Minerals:			
Calcium	0.85	0.90	0.95
Total phosphorus	0.65	0.70	0.75
Available phosphorus	0.42	0.45	0.48
Zinc (mg/kg)	60	60	60

*In the absence of experimental data, concentrations of trace elements and minerals may follow those for the starter turkey (*Table 53*).

Table 83 Recommended dietary energy levels and concentrations (%) of crude protein, amino acids and minerals for the quail during the finishing phase (21 days to slaughter)*

	Dietary energy level (kcal ME/kg)		
	2800	*3000*	*3200*
Crude protein	18.00	19.30	20.60
Amino acids:			
Lysine	1.15	1.23	1.31
Methionine	0.34	0.36	0.38
Sulphur amino acids	0.75	0.80	0.85
Tryptophan	0.18	0.19	0.20
Threonine	0.66	0.71	0.76
Glycine + serine	1.87	2.00	2.13
Leucine	1.13	1.21	1.29
Isoleucine	0.59	0.63	0.67
Valine	0.83	0.89	0.95
Histidine	0.45	0.48	0.51
Arginine	1.17	1.25	1.33
Phenylalanine + tyrosine	1.55	1.66	1.77
Minerals:			
Calcium	0.85	0.90	0.95
Total phosphorus	0.60	0.65	0.70
Available phosphorus	0.37	0.40	0.43
Zinc (mg/kg)	60	60	60

*For other constituents see *Table 53* (growing turkey).

Breeder quail

Laying performance of quail is very high when expressed in terms of body weight: thus egg mass per day approaches 9 g at a body weight of 175 g which, as a ratio to body weight, is double that of the laying chicken. In the quail, as with the majority of avian females, dietary energy level has little influence on rate of lay and therefore a fairly wide range of dietary energy concentrations may be used.

Daily protein and amino acid requirements for breeder quail are presented in *Table 84*. As a consequence of the high level of performance of this species coupled with a low feed intake (around 20 g/day) the dietary levels of protein and sulphur amino acids to be used are considerably higher than those with the laying chicken. Recommended amounts are given in *Table 85*.

Table 84 Daily energy and nutrient allowances for breeder quail

ME (kcal)	65
Crude protein (g)	4.5
Amino acids (g):	
Lysine	0.20
Methionine	0.09
Sulphur amino acids	0.16
Tryptophan	0.44
Threonine	0.14
Minerals (g):	
Calcium	0.73
Total phosphorus	0.15
Available phosphorus	0.09
Sodium	0.03
Chloride	0.03
Linoleic acid (g)	0.33

Table 85 Recommended dietary energy levels and concentrations (%) of crude protein, amino acids and minerals for breeder quail*

	Dietary energy level (kcal ME/kg)		
	260	280	300
Crude protein	17.8	19.2	20.6
Amino acids:			
Lysine	1.02	1.10	1.18
Methionine	0.38	0.41	0.44
Sulphur amino acids	0.72	0.78	0.84
Tryptophan	0.20	0.21	0.22
Threonine	0.54	0.58	0.62
Minerals:			
Calcium	3.00	3.20	3.40
Total phosphorus	0.60	0.65	0.70
Available phosphorus	0.37	0.40	0.43
Sodium	0.14	0.15	0.16
Chloride	0.13	0.14	0.15
Expected daily feed intake (g) at an ambient temperature of 23°C	25	23	21

*For other constituents *see Table 53* (breeder turkey).

Chapter 16
Nutrition of pheasants and partridges

Pheasants

With respect to behaviour and nutrient requirements, the growing pheasant is closer to the young turkey and quail than other avian species. If it is possible to adopt a dietary energy level between 2500 and 3000 kcal ME/kg, it is essential to provide high levels of protein during the first four weeks of life (*Tables 86* and *87*).

Table 86 Recommended dietary energy levels and concentrations (%) of crude protein, amino acids and minerals for the pheasant during the starter phase (0–4 weeks)*

	Dietary energy level (kcal ME/kg)			
	2500	*2700*	*2900*	*3100*
Crude protein	23.1	25.0	26.8	28.7
Amino acids:				
Lysine	1.30	1.40	1.50	1.60
Methionine	0.42	0.46	0.50	0.54
Sulphur amino acids	0.83	0.90	0.97	1.04
Tryptophan	0.20	0.22	0.23	0.24
Threonine	0.74	0.80	0.86	0.92
Minerals:				
Calcium	1.00	1.10	1.20	1.30
Total phosphorus	0.78	0.84	0.90	0.95
Available phosphorus	0.56	0.61	0.65	0.69

*In the absence of experimental data, concentrations of trace elements and vitamins may follow those for the starter turkey (*Table 53*).

Feed provided to pheasants for shooting must ensure that they are excellent fliers, which argues against the use of cereals alone. It is preferable to utilize a compound feed (*Table 88*) or a supplement to the cereals; in the latter case, the final mix offered should have similar characteristics to that of the former.

The adult hen pheasant does not have particularly high nutrient requirements. Either a diet designed for laying chickens or one with the characteristics shown in *Table 89* may be used.

Table 87 Recommended dietary energy levels and concentrations (%) of crude protein, amino acids and minerals for the pheasant during the growing phase (5–12 weeks)*

	Dietary energy level (kcal ME/kg)		
	2500	2700	2900
Crude protein	14.8	16.0	17.2
Amino acids:			
Lysine	0.79	0.85	0.91
Methionine	0.35	0.38	0.41
Sulphur amino acids	0.56	0.60	0.64
Tryptophan	0.13	0.14	0.15
Threonine	0.46	0.50	0.54
Minerals:			
Calcium	0.90	0.95	1.00
Total phosphorus	0.65	0.70	0.75
Available phosphorus	0.43	0.46	0.49

*For other constituents see Table 53 (growing turkey).

Table 88 Recommended dietary energy levels and concentrations (%) of crude protein, amino acids and minerals for pheasants (after 12 weeks)*

	Dietary energy level (kcal ME/kg)		
	2500	2700	2900
Crude protein	13.0	14.0	15.0
Amino acids:			
Lysine	0.60	0.65	0.70
Methionine	0.23	0.25	0.27
Sulphur amino acids	0.46	0.50	0.54
Tryptophan	0.11	0.12	0.13
Threonine	0.35	0.38	0.41
Minerals:			
Calcium	0.80	0.85	0.90
Total phosphorus	0.55	0.60	0.65
Available phosphorus	0.34	0.37	0.40

*For other constituents see Table 53 (finishing turkeys).

Partridges

Growing partridges have lower protein and amino acid requirements than growing pheasants. If it is possible, therefore, to feed them on diets designed for growing pheasants, levels may be reduced accordingly with no effect on growth, feathering or health status. Specific recommendations for growing partridges are given in Tables 90 and 91 and those during reproduction in Table 92.

Table 89 Recommended dietary concentrations (%) of crude protein, amino acids and minerals for the breeder pheasant*

	Dietary energy level (kcal ME/kg)		
	2500	*2700*	*2900*
Crude protein	12.5	13.5	14.5
Amino acids:			
Lysine	0.62	0.67	0.72
Methionine	0.27	0.29	0.31
Sulphur amino acids	0.47	0.51	0.55
Tryptophan	0.13	0.14	0.15
Threonine	0.41	0.45	0.48
Minerals:			
Calcium	2.40	2.60	2.80
Total phosphorus	0.54	0.57	0.60
Available phosphorus	0.30	0.32	0.34

*For other constituents *see Table 53* (breeder turkey).

Table 90 Recommended dietary concentrations (%) of crude protein, amino acids and minerals for the partridge during the starter phase (0–4 weeks)*

	Dietary energy level (kcal ME/kg)		
	2600	*2800*	*3000*
Crude protein	17.6	19.0	20.4
Amino acids:			
Lysine	1.02	1.10	1.18
Methionine	0.39	0.42	0.45
Sulphur amino acids	0.73	0.79	0.85
Tryptophan	0.18	0.19	0.20
Threonine	0.60	0.65	0.70
Minerals:			
Calcium	0.90	1.00	1.10
Total phosphorus	0.70	0.75	0.80
Available phosphorus	0.47	0.50	0.53

*In the absence of experimental data, concentrations of trace elements and vitamins may follow those for the starter turkey (*Table 53*).

Table 91 Recommended dietary concentrations (%) of crude protein, amino acids and minerals for the partridge during the growing phase (after 4 weeks)*

	Dietary energy level (kcal ME/kg)		
	2500	*2700*	*2900*
Crude protein	14.0	15.0	16.0
Amino acids:			
Lysine	0.74	0.80	0.86
Methionine	0.32	0.35	0.38
Sulphur amino acids	0.54	0.58	0.62
Tryptophan	0.13	0.14	0.15
Threonine	0.44	0.48	0.52
Minerals:			
Calcium	0.85	0.90	0.95
Total phosphorus	0.56	0.61	0.65
Available phosphorus	0.34	0.37	0.40

*For other constituents *see* Table 53 (growing turkey).

Table 92 Recommended dietary concentrations (%) of crude protein, amino acids and minerals for the breeder partridge*

	Dietary energy level (kcal ME/kg)		
	2600	*2800*	*3000*
Crude protein	14.7	16.0	17.0
Amino acids:			
Lysine	0.78	0.84	0.90
Methionine	0.33	0.35	0.38
Sulphur amino acids	0.61	0.66	0.70
Tryptophan	0.16	0.18	0.19
Threonine	0.49	0.53	0.57
Minerals:			
Calcium	2.40	2.60	2.80
Total phosphorus	0.54	0.57	0.60
Available phosphorus	0.30	0.32	0.34

*For other constituents *see* Table 53 (breeder turkey).

Part III
Composition of raw materials

Chapter 17
Comments on raw materials

Analytical values of raw materials
Principles

The tables in Chapter 18 only include those raw materials currently used in the feeding of pigs, rabbits and poultry; in addition to chemical measurements (e.g. protein level) they contain values appropriate to each species utilized (e.g. poultry metabolizable energy).

For the most commonly used raw materials, the most likely mean value together with the standard deviation is given to assist the feed formulator in deciding upon the most appropriate margin of security to use. In this context it may be recalled that the mean ±2 standard deviations will encompass 95% of all samples.

Among industrial by-products, those corresponding to the most widespread technological processes are included; more than one value is given if more than one process coexists with another.

A large proportion of the values arise from experiments carried out by INRA either working alone or in collaboration with other organizations (technical institutes or private companies). For the remainder, reviews of the literature have been used; in all cases comparisons have been made between results obtained by INRA and those previously published internationally.

Analytical techniques used

Measurements of dry matter levels are the most commonly used; a simple calculation allows a sample with a different dry matter content from the one cited to be characterized. This point is particularly important because, for a number of raw materials (cereals for example), variations in moisture content explain a large proportion of the differences observed for other characteristics (especially dietary energy values).

Current methods of analysis of raw materials rely frequently upon the Weende system: crude protein (N × 6.25), ether extract, crude fibre, ash and, by difference, nitrogen-free extract (NFE). The latter is defined therefore as the difference between total organic matter and the sum of protein, fat and crude fibre.

The deficiencies of this system are known: errors of overestimation of true protein (the most common), difference between true cellulose and crude fibre, accumulative errors with nitrogen-free extract which moreover encompasses carbohydrates with a variable nutritive value (starch, mono- and oligosaccharides, hemicelluloses, pectins).

The nutritional significance of the nitrogen-free extract varies according to the species in question; in poultry only starch and some mono- and oligosaccharides are utilizable. On the other hand, in pigs and, to a lesser extent, in rabbits an appreciable part of the other carbohydrate constituents (particularly hemicelluloses) may be digested. In rabbits a minimum amount of indigestible crude fibre is also recommended to prevent digestive disorders; this nutritional characteristic has been introduced into tables if sufficient information is available.

There are several methods for estimating starch which are not equivalent; enzymatic methods are the most specific whereas those based upon an acid hydrolysis (Ewers) overestimate starch by including certain non-amylaceous carbohydrates. Unfortunately, often only values obtained by the latter method are available.

Finally it is proposed that certain raw materials be characterized according to Van Soest: neutral detergent fibre (NDF) which corresponds mainly to the sum of cellulose, hemicellulose and lignin, and acid detergent fibre (ADF) which consists approximately of cellulose and lignin.

Energy values

Energy values supplied in the tables are all expressed in kcal/kg on a fresh weight basis. The most frequently cited figures are those referring to gross energy (heat of combustion), digestible energy for pigs and rabbits, and metabolizable energy values for pigs and poultry. Chapter 2 deals with definitions of terms and the choice of which system to use. With poultry specifically, ME values 'poussin' are those determined with birds less than 21 days of age which utilize some raw materials very poorly; they take into account a nitrogen retention of 40% of that ingested; ME 'broiler and layer' refer to those birds beyond 21 days of age and hens in lay and are based upon a nitrogen retention of 30%; ME 'adult cock' represents that value derived with adults at zero nitrogen retention and is that most frequently found in scientific publications. All poultry ME values are apparent.

Protein and amino acids

Levels provided in the current tables are those most commonly found expressed as crude protein (Kjeldahl; $N \times 6.25$).

Essential amino acid levels presented are mean and appropriate only to the figure for crude protein presented. As an initial approximation, it would be possible, simply from the figures given, to calculate the amino acid levels of raw materials where the crude protein figure was different from that provided. With some raw materials (wheat, maize, barley, peas and field beans) regression equations are available which allow a more precise estimation of the lysine level as a function of crude protein content to be determined. These equations, which are based on the assumption that the lysine level is not a constant function of crude protein, are given opposite:

Wheat: Lysine (% DM) = 0.0173 Crude protein (% DM) + 0.145
Maize: Lysine (% DM) = 0.0159 Crude protein (% DM) + 0.131
Barley: Lysine (% DM) = 0.0234 Crude protein (% DM) + 0.167
Peas: Lysine (% DM) = 0.0594 Crude protein (% DM) + 0.364
Field beans: Lysine (% DM) = 0.0467 Crude protein (% DM) + 0.530

Apparent protein digestibility is given for each species. Chapter 3 discusses definitions and describes methodology. Although there is much to be learnt concerning this subject, the values given do allow particularly good or bad sources of digestible protein to be identified.

Minerals and vitamins

Minerals

Chapter 4 presents the basic information relevant to minerals. Within the tables of raw material composition (*Table 93*) biologically available phosphorus values are only given for poultry with which a number of studies have been carried out. These values, determined with studies of bone mineralization in young chicks, often question most former values based upon a fixed availability of plant phosphorus of 30% and the incorrect assumption that phytate phosphorus is frequently unavailable.

Levels of trace elements in raw materials are presented separately in *Table 94*. These values are unlikely to be taken into account during feed formulation but they allow a better estimate of total trace element intake during nutrition experiments. It will be remembered however that the levels are frequently associated with considerable variability due in particular to the nature of the soil.

The columns 196 to 230 in *Table 93* concern mineral sources of feed-grade quality. *Table 95* presents the percentage composition of salts and oxides used in practice or during experiments studying the levels of macro and trace elements. The relative values on a weight basis of various minerals and their oxides, as the latter are those frequently used by the supplement industry, are given in *Table 96*.

Vitamins

It is usual and advisable to ignore vitamin A and D levels in raw materials and systematically to add to diets those amounts of synthetic vitamins corresponding to actual requirements for each age and species. Accordingly, only vitamin E and water-soluble vitamin levels are found in *Table 97*. Moreover, only those raw materials that could be properly considered as sources of vitamins are included. When considering mean values, it should always be remembered that vitamin content of raw materials may be very variable and that supplementation must consequently be considered.

Other characteristics

These concern those diverse substances which are essential or important from the point of view of animal performance and which may be deficient in current diets. This is the case with linoleic acid, a polyunsaturated fatty acid, which is necessary for all species at all ages but which in practice is only likely to be deficient for the laying hen. The raw materials therefore selected for inclusion in *Table 98* reflect this situation.

Carotenoid pigments (or xanthophylls) are also important in this context and they are necessary in the production of eggs, yellow broilers and guinea-fowl. These pigments are very unstable and oxidize at high temperatures. The mean levels indicated in *Table 99* are only relevant therefore to those raw materials that are prepared and stored properly. If there is any doubt, then it is advisable to check levels in the available raw materials.

Figures for the activity of carotenoids included in this table are only guidelines; they are only applicable to the level of transfer of pigments to the yolk of eggs and do not take into account either the colour obtained or that high concentrations reduce this level of transfer with certain carotenoids. Nevertheless it may be stated that yolk colour is directly influenced by this activity, when the colour intensity wished for is not higher than 9 on the Roche fan.

Advantages and disadvantages of principal raw materials

The nutritive value of a raw material is not limited to its proximate analysis. The presence of antinutritional or toxic factors are among those that may influence, to a greater or lesser extent, the ease of digestibility of major nutrients. These factors in the true sense are those that, even in trace amounts, reduce or completely prevent the utilization of a nutrient either at the digestive or metabolic level. The best known are antivitamin factors, organic acids which chelate mineral cations and antienzyme agents (antitrypsins in soya). Condensed tannins, lectins, saponins and certain alkaloids are, equally, substances which can be classed as antinutritive.

In addition other undesirable products may be present and may act in one or more of the following ways:

(1) actual toxicity (e.g. mycotoxins);
(2) reduction in appetite (e.g. isothyocyanates in rapeseed);
(3) gastroenteritis (e.g. α-galactosides);
(4) alteration of organoleptic (e.g. trimethylamine) or physical qualities of the product.

All of these possibilities will now be considered, relative to the principal classes of raw materials.

Cereals

Among the common cereals, *maize* has the highest dietary energy value as a consequence of its high level of starch and oil (4%). This quality is of particular value to poultry and also to an extent to pigs. On the other hand the low protein levels, and the relatively poor content of lysine and tryptophan within the protein, must be underlined; this problem is partially overcome in poultry because of a high digestibility. Maize is a poor source of certain trace elements and vitamins (available niacin) but has useful amounts of biotin and carotenoids. Phosphorus present is fairly available to poultry because of phytase activity within the grain.

Maize has the considerable advantage of being of reasonably constant composition; its dietary energy value varies little, if the moisture content of the grain is taken into account, and only harvesting when not mature slightly lowers it.

Sometimes the accidental presence of fungal toxins, principally zearalenone, aflatoxin and ochratoxin, has been established. Harvesting under damp conditions and storage in open stores are those factors which contribute most to the development of the microorganisms responsible, particularly *Fusarium* which produces zearalenone, known above all for the adverse effects it may have on sow productivity when present at levels above 3 mg/kg.

As an excellent source of linoleic acid, the use of maize may contribute to the production of fat depots high in polyunsaturated fatty acids.

Soft wheat is, surprisingly, only infrequently used in animal nutrition in France. While it is true that because of its low content of pigments it is not appropriate for the prodution of eggs or yellow broilers, it is on the other hand perfectly suited for all other production systems. As with maize, its dietary energy value is relatively constant and not appreciably influenced by season, variety or growing conditions. Its content of protein which is highly digestible in pigs, is greater than that of maize but in addition more variable; a late application of nitrogen fertilizer coupled with the presence of poorly-formed grains during hot summers are particularly associated with higher levels.

The availability of phosphorus in wheat to poultry and pigs is good (equal to or higher than 50%) as a consequence of endogenous phytases; on the other hand biotin in wheat is unavailable to poultry. Finally the presence of wheat in a compound diet improves pellet quality.

Traditionally *barley* is fed to pigs and rabbits but not often to poultry. In the former, barley has the additional advantage of being palatable and its dietary energy value is as expected for an oil content of less than 2%; diets based on barley therefore are associated with the production of saturated carcass fat. In poultry the utilization of dietary energy is less efficient and high levels of barley (above 30%) may promote a lowering of performance in the young broiler. On the other hand, pullets and laying hens are able to tolerate high dietary levels of barley without any problems except that, as with wheat, the addition of carotenoids must be undertaken to obtain adequate yolk colour.

There are a number of different types of barley. Winter six-row types are higher in cellulose and have a lower dietary energy value than two-row winter and spring varieties whose nutritive values approach that of wheat. It is important to note that feed barley is frequently composed of small grains that have been rejected by maltsters; as a consequence these have the lowest dietary energy values but the highest protein contents.

It is possible to predict the digestible energy (DE) value of barley for pigs from the crude fibre (CF) content:

$$DE \text{ (kcal/kg DM)} = 4072 - 110 \text{ CF (\% DM)}$$

Sorghum is, by virtue of its composition, very close to maize. Bird-resistant varieties contain significant amounts of condensed tannins which lower protein digestibility and reduce dietary energy values for poultry and pigs. Accordingly two groups of figures are presented in *Table 93*, based upon the range of tannin contents likely to be found.

With respect to other cereals, *rye* is frequently only tolerated at low inclusion levels (*Table 101*) because of the content of β-glucans and phenolic products (N-alkyl resorcinol) which are associated with toxicity in poultry. There are many varieties of *triticale* (a hybrid between soft or hard wheat and rye); French types are very similar to soft wheat.

Cereal by-products

Wheat by-products (middlings, shorts and bran) are frequently used in pig feeding whereas those from maize, barley and rice are less commonly employed. When compared with the cereals from which they were derived, these raw materials have higher levels of protein and a better essential amino acid balance.

Middlings are rich in starch (20–58%) and oil (3.5–4%) and constitute therefore an excellent source of dietary energy which may without problems replace the majority if not all the cereals in a diet. They are incorporated up to 30–40% into diets for fattening pigs.

Shorts, by-products of flour mills manufacturing semolina, have levels of starch approaching that of middlings (23–40%) but a higher crude fibre content (4–8%). It is possible to distinguish between *white shorts* (up to 4% crude fibre) and *brown shorts* (7–8% crude fibre). These two categories of raw material may partially replace cereals in pig diets.

Among cereal by-products, *bran* represents that with the highest crude fibre content (9–10%) and, consequently, the lowest dietary energy value. As a result of its high content of cell-wall constituents (cellulose, hemicelluloses) it is an important source of bulk for breeding sows, associated with a not inconsiderable level of protein. Finely-ground bran is more digestible than when coarsely ground.

Ensiled cereals

This group is only of interest to the feeding of pigs. Figures reported in *Table 93* are concerned with *maize silages* when they are produced under practical conditions. Thus the level of maturity required when the entire plant is harvested is lower than when the cobs alone, entire or not, are used which is itself lower than that level of maturity when the grain is at the milky-white stage. To be of good quality silage, whatever the type, must be made rapidly, well compressed and protected from all contact with air.

The limit of incorporation of maize silage into a diet is inversely proportional to its crude fibre content and increases with the age of the animal. Thus that based on grain may be used at all ages whereas that from a mixture of grain and husks may only practically be used with growing pigs and beyond. The most fibrous silages may only be included up to a level of 20% in diets for young pigs but may contribute up to half the dietary energy of diets for sows.

The dietary energy value of maize silage, whatever type it may be, is more or less the same as the corresponding dried product. On the other hand protein digestibility is generally higher by one or two percentage points in the former.

Barley and wheat may equally be ensiled wet at a varying degree of maturity. The value of the products obtained is however frequently lower than the equivalent dried product, in particular for whole grains.

Molasses

Molasses are very low in protein and are energy feeds of variable nutritive value depending essentially upon the content of available carbohydrates. The major disadvantage is an excessive level of potassium and fermentable carbohydrates. The use of molasses in diets at significant levels is therefore accompanied by a reduction in digestibility and the appearance of diarrhoea; these problems limit its use in diets for pigs and poultry. Cane and sugar beet molasses have similar compositions, except for a higher potassium level in the latter.

Dietary oils and fats

The dietary energy value of oils and fats is influenced considerably by the level of saturated fatty acids palmitic and stearic; when the level exceeds 50%, the value decreases considerably. This is the case with *tallow* which is only moderately utilized by very young animals, in particular chicks below two weeks of age.

Vegetable oils and *poultry fat* are rich in unsaturated fatty acids and have high dietary energy values for all non-ruminant species. Moreover their addition to saturated fats (tallow) improves the value of the latter.

Finally the presence of an extra-caloric effect with poultry must be noted: frequently the incorporation of oils and fats into a feed improves its dietary energy value beyond that which is predictable from simple addition of individual energy values. Fats and oils therefore have a beneficial effect upon the utilization of other constituents of the diet.

Roots and tubers

Beets, either fresh or ensiled, are palatable to pigs. The level of water and bulkiness limits their level of incorporation into diets at the beginning of the growing period to 20% on a dry matter basis; this level may on the other hand progressively approach 60% with adults. In addition protein levels are poor both quantitatively and qualitatively. In a dried form, they may replace around a third of the cereal content of a diet.

Beet pulps, fresh or dried, have a high level of cellulose which limits their use; digestibility of their nutrients is good.

Potatoes contain starch which is very poorly digested by pigs. They must be cooked therefore to be efficiently utilized. Potatoes may be consumed immediately or ensiled but their bulkiness limits their level of incorporation into diets to 40% on a dry matter basis. This limit applies equally to certain by-products of the potato industry (puree); it must be reduced if the by-products contain sizeable amounts of non-starch material. Raw but ensiled potatoes may only be used at low dietary levels (below 20% on a dry matter basis) with growing pigs and following a short period of adaptation.

Dried potato granules are useful sources of dietary energy and may be included up to a level of 40% of the feed of pigs.

Manioc is very low in protein and is essentially an energy-yielding raw material high in available carbohydrates (60–75% starch). Its quality is very changeable as a consequence of variability in levels of crude fibre and ash (particularly silica) which may considerably reduce dietary energy values. It is possible to predict the digestible energy (DE) value of manioc from its crude fibre (CF, between 3.5 and 7.5%) and ash (between 1.7 and 8.1%) content with the following equation adopted from Dutch work:

$$DE \text{ (kcal/kg DM)} = 4400 - 110 \text{ CF (\% DM)} - 43 \text{ ASH (\% DM)}$$

Other adverse factors such as microbiological contamination and, to a lesser extent, the presence of cyanogenetic compounds may be found. Given the wide variability in their content between different samples of manioc, the limits of use given in *Tables 100* and *101* are only guidelines and must be modified in the light of the determined characteristics of products.

Various industrial by-products

Coffee grounds are characterized by very high levels of cell wall-related constituents, associated with prolonged cooking; this considerably reduces palatability and digestibility in pigs.

Apple pomace is rich in pectin and lignin and is only slightly digestible. Nevertheless its nutritive value is probably underestimated as the level of hemicelluloses is low or zero.

Residues from the fermentation of carbohydrates and alcohol distillation, *vinasses*, include under the same name a large number of different products which are characterized by their origin (beet or cane) and by their composition. The distinction made in the tables is based upon crude protein levels and mineral contents which are negatively correlated with each other. A large proportion of nitrogen in vinasses is non-protein, either ammoniacal or glutamic acid. Their high mineral levels are due above all to potassium and this limits their level of incorporation to between 5 and 10% of diets for fattening pigs. This figure may approach 10 to 15% in the case of vinasses which have had the potassium removed.

Oil meals

Soya-bean meal is particularly well suited to use in diets for non-ruminants. If not cooked it contains numerous very potent antinutritional factors which have to be removed by heat treatment. On the other hand overcooking may to an extent render amino acids unavailable. Soya-bean meal is relatively deficient in sulphur amino acids and available zinc. Phosphorus, similarly, is only sparingly available.

Sunflowerseed meal has only a moderate dietary energy value; the protein content is highly digestible but is deficient in lysine. The major problem with this raw material is its variable composition due to harvesting conditions and treatment during oil extraction. If deficiencies in dietary energy value and lysine are made up, it is an excellent raw material.

Rapeseed meal from *Brassica napus* and *B. campestris* contains protein that is unquestionably well-balanced. Unfortunately there are well-known disadvantages based, among others, upon the presence of glucosinolates which under the action of myrosinase produce compounds that are goitrogenic (oxazolidine thione) or bitter-tasting (isothyocyanate). As a result, the use of high levels of rapeseed meal is associated with a reduction in growth rate with all species and problems of inappetance in mammals (particularly pigs). In addition rapeseed meal increases mortality in laying hens because of haemorrhagic fatty liver. Finally the presence of sinapine promotes the accumulation of trimethylamine in the yolks of brown eggs and in some poultry meat which is associated with fishy taints if the level of incorporation of rapeseed meal is higher than 5%. The use of those meals low in thioglucosinolates (from 'double zero' varieties) reduces considerably the problems described above except those related to sinapine.

Rapeseed meal is also well utilized by pigs when considering both energy and protein; on the other hand protein and above all carbohydrates are poorly utilized in poultry.

The composition of rapeseed meal is very variable particularly with respect to its residual oil content and, consequently, its dietary energy value.

Plant proteins

As a consequence of a moderate protein level, *peas* are rarely considered suitable for feeding to young poultry during the starter phase. They are more readily incorporated into diets for finishing stock, breeders of all species and laying hens. Under these conditions the use of peas poses no problems as long as their level of inclusion does not exceed 20%.

Lupins are very acceptable to poultry both old and young. They are deficient in lysine, sulphur amino acids and tryptophan but extremely rich in manganese without this apparently presenting any problems. The addition of lupins to diets increases folic acid requirements (by 0.02 mg/kg per 1% lupins) of the growing broiler. In pigs the intestinal fermentation of α-galactosides present in lupins restricts their level of incorporation (*Table 100*).

Field beans are well tolerated by pigs and poultry. Brown varieties (rich in tannins) are generally of less interest in the nutrition of these two; they have a lower dietary energy value and poorer protein digestibility. Tannins are not the only compounds responsible for these problems; vicin and convicin are equally of importance. Similarly tryptophan availability is lower in brown varieties.

Unicellular protein and algae

Yeasts may be classified into two major categories depending upon whether or not they have been specifically cultivated for animal nutrition. Yeasts from other industries (brewing, distilling) may be of very variable quality because it is the optimization of the quality of the major product rather than the yeast that is of importance. As a consequence of this variability and the risks of residues (alcohol) it is advisable to limit their use. On the other hand, feed yeasts of given origin specially grown as a protein source have a constant composition and are free of toxic residues. They may be used in the feeding of pigs as the sole complement to cereals if their price permits it. The high content of nucleic acids (5–12% of total nitrogen) in yeasts does not pose any particular problem for pigs that are able to dispose of them without any significant increase in uric acid levels.

Bacterial proteins (ICI Pruteen for example) are, like yeasts, rich in protein and nucleic acids; they are tolerated equally by both weaners and by growing/finishing pigs up to 15% and 10% respectively of the diet on a dry matter basis.

With growing birds, yeasts and bacterial proteins have in general a beneficial effect at low dietary levels (less than 5%) but higher amounts may be detrimental; the former (maximum level of 10%) are less well tolerated than the latter (maximum level of 15%). Both are better accepted by adult females (up to 20% of the diet).

Algae have a greater variability in composition than either yeasts or bacterial proteins. Those which appear to have a future, especially when associated with the treatment of waste waters, are *Chlorella* and *Scenedesmus* of which many varieties and types exist. They are frequently rich in pigments and cell wall-related compounds, and may be incorporated into diets for pigs up to a level of around 10% on a dry matter basis.

Animal meals

This group of raw materials includes those by-products from various operations: fishing, meat preservation, slaughterhouses and other meat-related industries. These products are therefore very different from each other more because of the heterogeneity of the raw materials used than the means of processing employed. However it must be mentioned that there has been a considerable effort over the last few years to standardize the products, which has resulted in more constant and better-known analytical characteristics. This progress has been taken into account and the values presented in *Table 93* are for those meals most commonly encountered. However, there is variability in animal meals which means that equations predicting their dietary energy value need to be derived. Some equations applicable to poultry are presented in the following paragraphs where their accuracy has been verified.

Fish meals are excellent sources of protein and minerals. Their dietary energy value varies above all as a function of their residual oil content and mineral levels. The equation of Opstvedt, based on crude protein (CP) and oil (EE), which relates to adult cock metabolizable energy (ME) values may therefore be used:

$$ME(kcal/kg\ DM) = 39.5\ CP\ (\%\ DM) + 64.5\ EE\ (\%\ DM)$$

Meat meals are similarly separated from each other by their ash, crude protein and ether extract (EE) contents. Metabolizable energy (ME) values for adult cocks are predicted accurately using the following equation (Lessire and Leclercq):

$$ME(kcal/kg\ DM) = 3570 + 60\ EE\ (\%\ DM) - 45.5\ ASH\ (\%\ DM)$$

Phosphorus in meat meals has a high availability but not as high as that from dicalcium phosphate. Their high calcium levels, moreover, limit the level of inclusion in diets for growing animals (*Tables 100* and *101*).

Milk by-products

This group is relevant only to pigs.

Skimmed milk and *buttermilk* are balanced protein raw materials that may constitute the sole protein complement to cereal-based diets for pigs of any age. In small amounts they are essential in weaner diets. When dried, their level of inclusion in pelleted diets must be restricted in order to avoid pellets that are too hard.

Wheys are very variable and may be classified with reference to the different technological processes involved in their manufacture. Three main types exist: *sweet* (from curdling under pressure), *acid* (from lactic acid curdling or the production of casein with mineral acids) and *mixed* according to whether the former or latter predominates. They are essentially energy-yielding raw materials, although their protein fractions are better balanced than those of cereals. Their nutritive value and theoretical limits of inclusion are relatively constant whatever the form, being liquid, concentrated or powdered. However, powdered whey presents technological problems as it is hydroscopic and pellets in which it is contained rapidly harden. From a nutritional point of view lactose, the value of which falls with age, is the principal factor that limits the incorporation of wheys into diets for growing pigs. The mean limit is 30% of the dry matter. This may be altered as a function of age of the animal and economic considerations.

Ultrafiltrates from milk and whey, together with *deproteinized serums* are simply solutions of lactose and minerals. Their level of incorporation must be limited to a level of 20% or less of the dry matter of the diet.

Delactosed serums are products that still contain 50% lactose but with low protein levels because of the technological processes to which they have been subjected.

Recommended limits of incorporation of some raw materials into diets for pigs and poultry are included in *Tables 100* and *101*. These limits are based upon experience within INRA and many observations from elsewhere. However they must only be regarded as guidelines and are not definitive insofar as the genetic selection of plants and the processing of compound diets are likely gradually to reduce the antinutritional factors still present in current raw materials.

In addition to these guidelines on the use of raw materials, legislation governing the maximum permissible levels of certain antinutritive substances is given in *Table 102*.

Chapter 18
Tables of raw material composition

Table 93 includes essential characteristics used in diet formulation. Raw materials included are classified into 16 principal groups and each has a code number which appears within the tables and in an index at the end of the book.

All values presented are on a fresh weight basis.

Other tables which follow are:

Table 94   Trace element content of the raw materials.
Table 95   Basic composition of the main mineral salts and oxides.
Table 96   Equivalence factors between minerals and their oxides.
Table 97   Vitamin content of major raw materials.
Table 98   Linoleic acid content of major raw materials.
Table 99   Xanthophyll content of major raw materials.
Table 100 Inclusion limits of some raw materials: pigs.
Table 101 Inclusion limits of some raw materials: poultry.
Table 102 Maximum permissible levels of certain non-nutritive factors.

Table 93 Raw material composition

Code		Raw material	Page
	I	**Major cereals**	
1		Oats	141
2		Wheat — soft	141
3		Maize (corn, USA)	141
4		Barley — 2 row	141
5		Barley — 6 row	142
6		Sorghum — low tannin	142
7		Sorghum — high tannin	142
	II	**Minor cereals**	
8		Oats — naked	143
9		Oats — dehulled	143
10		Oats — flaked	143
11		Wheat — hard	143
12		Millet	143
13		Barley — naked	144
14		Rice	144
15		Buckwheat	144
16		Rye — winter	144
17		Triticale — French	144
	III	**Cereal by-products**	
		From hard wheat	
18		Middlings	145
19		Shorts	145
20		Bran	145
		From soft wheat	
21		Middlings	145
22		Germ	145
23		White shorts	146
24		Red shorts	146
25		Fine bran	146
26		Coarse bran	146
		From maize (corn, USA)	
27		Distiller's dried grains and solubles (DDGS)	147
28		Germ	147
29		Gluten feed	147
30		Gluten meal 40	147
31		Gluten meal 60	147
32		Cobs	147
33		Bran	148
34		Germ — oilmeal (expeller)	148
35		Germ — oilmeal (solvent)	148
		From barley	
36		Brewer's grains	148
37		Rootlets	149
		From rice	
38		Broken	149
39		Shorts	149
40		Bran	149

Table 93 (cont.)

Code	Raw material	Page
	IV Ensiled cereals	
	Maize (corn, USA)	
41	Ears and stalk tops	150
42	Ears and husks	150
43	Ears without husks	150
44	Grains	150
45	Whole plant	150
	V Carbohydrate raw materials	
46	Maize starch	151
47	Ensiled unripe bananas	151
48	Whole — unripe bananas	151
49	Whole — ripe bananas	151
50	Carob fruit	151
51	Carob germ	152
52	Dehulled chestnut	152
53	Whole chestnut	152
54	Breadfruit	152
55	Dehulled acorns	152
56	Whole acorns	153
57	Dasheen	153
58	Sugar beet molasses	153
59	Sugar cane molasses	154
60	Apples	154
61	Sugar	154
	VI Fats/oils	
62	Animal fat	155
63	Poultry fat	155
64	Vegetable oil	155
65	Lard	155
66	Tallow	155
	VII Roots and tubers	
67	Sugar beet	156
68	Fodder sugar beet	156
69	Fodder beet	156
70	Sugar beet pulp	156
71	Arrowroot	156
72	Carrots	157
73	Chicory leaves	157
74	Chicory roots	157
75	Yam	157
76	Cassava — pelleted	157
77	Cassava roots	158
78	Turnips	158
79	Sweet potato	158
80	Potato starch	159
81	Potatoes — raw	159

Table 93 (cont.)

Code		Raw material	Page
82		Potatoes — dried	159
83		Potato starch flour	159
84		Potato flakes	160
85		Potato proteins	160
86		Potato pulp	160
87		Swedes	160
88		Jerusalem artichoke	160
	VIII	**Industrial by-products**	
89		Dried citrus pulp	161
90		Coffee grounds	161
91		Apple pomace	161
92		Grape pips	161
93		Grape pulp	162
94		Grapeseed oil meal	162
95		Tomato skins	162
96		Tomato pips	162
97		Tomato pulp	163
98		Vinasse	163
99		Vinasse — sugar waste	163
100		Vinasse — yeast waste	163
	IX	**Various vegetable products**	
101		Cocoa hulls	164
102		Rapeseed hulls	164
103		Cabbages	165
104		Brussels sprouts	165
105		Cauliflowers	166
106		Marrowstem kale	166
107		Cereal straw	166
108		Grass meal	166
109		Dehydrated Lucerne meal 21	167
110		Dehydrated Lucerne meal 17	167
111		Dehydrated Lucerne meal 15	167
112		Dehydrated Lucerne meal 12	167
113		Lucerne protein concentrate	168
114		Soya-bean hulls	168
	X	**Vegetable proteins**	
115		Full fat rapeseed	169
116		Field beans	169
117		Toasted beans	169
118		Lentils	170
119		Sweet white lupins	170
120		Smooth winter peas	170
121		Smooth spring peas	171
122		Full fat soya-beans	171
123		Soya protein isolate	171

Table 93 (*cont.*)

Code	Raw material	Page
	XI Oil meals	
124	Peanut meal	172
125	Dehulled rapeseed meal	172
126	Rapeseed meal — expeller	172
127	Rapeseed meal — solvent-extracted	172
128	Copra meal	173
129	Cotton seed meal	173
130	Palm kernel meal	173
131	Soya-bean meal 44	174
132	Soya-bean meal 48	174
133	Soya-bean meal 50	175
134	Sunflower seed meal	175
	XII Microbial proteins/algae	
135	Algae — Spirulina	176
136	Algae — Chlorella	176
137	Algae — Scenedesmus	176
138	Brewer's yeast	176
139	Distiller's yeast	177
140	Torula yeast	177
141	Lactic yeast	177
142	Fodder protein	177
143	ICI Pruteen	177
	XIII Animal by-products	
144	Fish protein concentrate	178
145	Krill concentrate	178
146	Krill meal	178
147	Fish meal 60% crude protein	179
148	Fish meal 65% crude protein	179
149	Fish meal 72% crude protein	179
150	Fish meal defatted, 65% crude protein	179
151	Fish meal defatted, 72% crude protein	179
152	Fish solubles	180
153	Greaves	180
154	Feather meal	180
155	Blood meal	180
156	Meat meal 50% crude protein	180
157	Meat meal 55% crude protein	181
158	Meat meal defatted, 50% crude protein	181
159	Meat meal defatted, 55% crude protein	181
160	Meat meal defatted, 60% crude protein	181
161	Meat and bone meal	182
162	Poultry litter	182
163	Poultry offal meal	182
164	Poultry hatchery waste	182
	XIV Dairy products	
	Liquid	
165	Milk	183
166	Skimmed milk	183

Table 93 (cont.)

Code		Raw material	Page
167		Sweet buttermilk	183
168		Sweet separated whey	183
169		Acid separated whey	183
170		Mixed separated whey	184
171		Acid goat whey	184
172		Whey — rennet casein	184
173		Whey — lactic acid casein	184
		Dried	
174		Milk	185
175		Skimmed milk	185
176		Sweet buttermilk	185
177		Lactic acid casein	185
178		Hydrochloric acid casein	185
179		Sodium caseinate	186
180		Sweet separated whey	186
181		Mixed separated whey	186
182		Sweet ewe whey	186
183		Milk permeate	187
184		Mixed whey permeate	187
185		Partially delactosed whey	187
186		Whey proteins	187
	XV	**Amino acids**	
187		*dl*-Methionine	188
188		*l*-Lysine hydrochloride	188
189		Methionine hydroxyanalogue	188
	XVI	**Minerals**	
		Natural sources (feed grade)*	
		Sources of calcium	
190		Calcium carbonate	189
191		Limestone	189
192		Sea shells	189
193		Carbonate from sugar beet factories	189
194		Oyster shells	189
195		Sea shells	189
196		Dried egg shells	189
197		Cleaned egg shells	189
198		Marl	189
199		Hatchery ashes	189
200		Portland cement	189
201		Gypsum	189
		Sources of water-soluble phosphorus	
202		Phosphoric acid	190
203		Monosodium phosphate dihydrate	190
204		Anhydrous monosodium phosphate	190
205		Disodium phosphate dihydrate	190
206		Anhydrous disodium phosphate	190
207		Monopotassium phosphate	190

Table 93 *(cont.)*

Code	Raw material	Page
208	Dipotassium phosphate	190
209	Monoammonium phosphate	190
210	Diammonium phosphate	190
211	Sodium tripolyphosphate	190
212	Monocalcium phosphate	190
	Sources of sparingly soluble or insoluble phosphorus	
213	Hydrated dicalcium phosphate	191
214	Anhydrous dicalcium phosphate	191
215	Mono/dicalcium phosphate	191
216	Tricalcium phosphate	191
217	Rock phosphate	191
218	Defluorinated rock phosphate	191
219	Bone meal	191
220	Degelatinized bone meal	191
221	Ca-Mg-Na phosphate	191
	Sources of sodium	
222	Sodium chloride	191
223	Sea salt	191
224	Sodium bicarbonate	191

*For reagent-grade products *see* Table 95.

Table 93 (cont.)

Major cereals	Code:	1	2		3		4	
Characteristics (% fresh weight)		Oats	Soft wheat		Maize		Barley 2 row	
			Mean	SE*	Mean	SE	Mean	SE
Dry matter		86	86		86		86	
Gross energy (kcal/kg)		4 010	3 790	55	3 860	63	3 780	40
Nitrogen-free extract		57.8	68.8		69.0		67.3	
Starch–acid hydrolysis		37.0	56.0	2.4	60.5	1.9	50.5	2.5
Starch–enzymic hydrolysis		34.0					49.0	
Sugars					2.1			
Ether extract		5.3	1.9	0.4	4.2	0.5	2.0	0.2
Crude fibre		10.2	2.3	0.3	2.2		4.4	0.6
Acid detergent fibre (ADF)		13.0	3.3		3.0		5.5	1.0
Neutral detergent figre (NDF)		26.0	10.5		9.0		15.0	1.5
Crude protein		**10.0**	**11.3**	**0.8**	**9.0**	**0.8**	**10.0**	**1.1**
Amino acids:								
Lysine		0.40	0.32		0.25		0.37	
Methionine		0.16	0.19		0.19		0.17	
Methionine + cystine		0.50	0.47		0.39		0.42	
Tryptophan		0.12	0.13		0.06		0.11	
Threonine		0.35	0.34		0.32		0.34	
Glycine + serine		1.00	1.00		0.78		0.81	
Leucine		0.73	0.76		1.13		0.70	
Isoleucine		0.42	0.42		0.35		0.38	
Valine		0.52	0.54		0.46		0.53	
Histidine		0.22	0.26		0.26		0.22	
Arginine		0.66	0.54		0.43		0.50	
Phenylalanine + tyrosine		0.90	0.82		0.85		0.86	
Ash		**2.70**	**1.65**	**0.30**	**1.35**	**0.18**	**2.30**	**0.21**
Calcium		0.08	0.06		0.01		0.05	
Total phosphorus		0.34	0.33		0.27		0.35	
Sodium		0.07	0.05		0.01		0.05	
Potassium		0.42	0.40		0.33		0.48	
Chloride		0.10	0.06		0.05		0.14	
Magnesium		0.14	0.12		0.11		0.12	
Pig								
Digestible energy — DE (kcal/kg)		2 730	3 310		3 400		3 020	
Metabolizable energy — ME (kcal/kg)		2 650	3 210		3 315		2 935	
Apparent crude protein digestibility — ACPD (%)		79	87		81		78	
Rabbit								
Digestible energy — DE (kcal/kg)		2 800	3 070		3 260		3 000	
Indigestible crude fibre — IDCF (%)		9.8	1.0		0.6		3.8	
Apparent crude protein digestibility — ACPD (%)		80	83		82		72	
Poultry								
Metabolizable energy — ME (kcal/kg):								
Broiler			2 850	70	3 200	65	2 400	90
Pullet and laying hen		2 560	3 050	72	3 300	65	2 840	90
Adult bird		2 520	2 995	72	3 250	65	2 790	90
Apparent crude protein digestibility — ACPD (%)			82		86		70	
Available phosphorus (%)**		**0.08**	**0.18**		**0.05**		**0.17**	

*SE = standard error.
**In the majority of cases figures are determined; for others vegetable phosphorus is assumed to be 30% available.

Table 93 (cont.)

Major cereals Characteristics (% fresh weight)	Code:	5 Barley 6 row		6 Sorghum Low tannin		7 Sorghum High tannin	
		Mean	SE*	Mean	SE	Mean	SE
Dry matter		86		86		86	
Gross energy (kcal/kg)		3 770	40	3 820	50	3 820	50
Nitrogen-free extract		67.1		69.0		68.3	
Starch–acid hydrolysis		49.5	1.9	59.8	2.6	56.8	2.6
Starch–enzymic hydrolysis		44.0					
Sugars							
Ether extract		1.8	0.3	3.0	0.2	3.0	0.2
Crude fibre		5.6	0.7	2.5	0.1	3.0	0.1
Acid detergent fibre (ADF)		6.5	1.0	3.8		4.6	
Neutral detergent figre (NDF)		18.0	1.5	9.0		9.8	
Crude protein		**9.2**	**0.9**	**10.0**	**1.4**	**10.0**	**1.4**
Amino acids:							
Lysine		0.35		0.23		0.23	
Methionine		0.16		0.16		0.16	
Methionine + cystine		0.41		0.33		0.33	
Tryptophan		0.10		0.09		0.09	
Threonine		0.31		0.33		0.33	
Glycine + serine		0.78		0.65		0.65	
Leucine		0.64		1.38		1.38	
Isoleucine		0.35		0.44		0.44	
Valine		0.50		0.55		0.55	
Histidine		0.20		0.22		0.22	
Arginine		0.48		0.39		0.39	
Phenylalanine + tyrosine		0.79		0.95		0.95	
Ash		**2.30**	**0.20**	**1.45**	**0.10**	**1.65**	**0.10**
Calcium		0.05		0.03		0.04	
Total phosphorus		0.36		0.30		0.34	
Sodium		0.04		0.01		0.01	
Potassium		0.44		0.35		0.37	
Chloride		0.14		0.10		0.11	
Magnesium		0.12		0.15		0.16	
Pig							
Digestible energy — DE (kcal/kg)		2 890		3 200		3 060	
Metabolizable energy — ME (kcal/kg)		2 810		3 115		2 980	
Apparent crude protein digestibility — ACPD (%)		76		75		68	
Rabbit							
Digestible energy — DE (kcal/kg)		2 950		3 200		3 160	
Indigestible crude fibre — IDCF (%)		4.1		0.1		0.8	
Apparent crude protein digestibility — ACPD (%)		72		84		81	
Poultry							
Metabolizable energy — ME (kcal/kg):							
Broiler		2 300	87				
Pullet and laying hen		2 745	87	3 180	60	2 835	80
Adult bird		2 695	87	3 140	60	2 796	80
Apparent crude protein digestibility — ACPD (%)		70		75			
Available phosphorus (%)		**0.17**		**0.05**		**0.05**	

Table 93 (cont.)

Minor cereals	Code:	8	9	10	11	12
Characteristics (% fresh weight)		Oats			Wheat — hard	Millet
		Naked	Dehulled	Flaked		
Dry matter		86	86	88	87	88
Gross energy (kcal/kg)		3 960	4 110	4 180	3 880	3 950
Nitrogen-free extract		64.0	61.4	62.8	67.0	59.8
Starch–acid hydrolysis					53.5	52.1
Starch–enzymic hydrolysis		43.5				
Sugars					3.1	1.3
Ether extract		5.0	7.4	7.2	2.0	3.7
Crude fibre		2.0	1.8	2.0	2.5	9.3
Acid detergent fibre (ADF)		3.1	2.2		3.3	13.7
Neutral detergent figre (NDF)		7.9	9.2	10.4	16.1	
Crude protein		**13.0**	**12.9**	**13.6**	**13.7**	**11.5**
Amino acids:						
Lysine		0.53	0.50	0.52	0.37	0.19
Methionine		0.21	0.20	0.22	0.22	0.49
Methionine + cystine		0.65	0.64	0.68	0.56	0.68
Tryptophan		0.16	0.18	0.19	0.16	0.16
Threonine		0.44	0.41	0.46	0.38	0.44
Glycine + serine		1.25	1.17	1.40	1.18	0.90
Leucine		0.95	0.93	1.02	0.91	1.26
Isoleucine		0.53	0.53	0.58	0.52	0.43
Valine		0.68	0.70	0.76	0.60	0.54
Histidine		0.28	0.28	0.27	0.31	0.16
Arginine		0.88	0.79	0.91	0.64	0.62
Phenylalanine + tyrosine		1.17	1.04	1.25	1.05	1.01
Ash		**2.02**	**1.94**	**2.60**	**1.78**	**3.73**
Calcium		0.08	0.08	0.08	0.04	0.04
Total phosphorus		0.36	0.38	0.47	0.34	0.28
Sodium			0.06	0.01	0.01	0.04
Potassium			0.35	0.37	0.40	0.41
Chloride			0.05	0.04	0.05	0.03
Magnesium			0.12	0.17	0.10	0.15
Pig						
Digestible energy — DE (kcal/kg)		3 480	3 630	3 680	3 390	2 870
Metabolizable energy — ME (kcal/kg)		3 365	3 515	3 560	3 275	2 780
Apparent crude protein digestibility — ACPD (%)		85	85	85	86	74
Rabbit						
Digestible energy — DE (kcal/kg)		3 400				
Indigestible crude fibre — IDCF (%)		1.8				
Apparent crude protein digestibility — ACPD (%)		89				
Poultry						
Metabolizable energy — ME (kcal/kg):						
Broiler					2 980	
Pullet and laying hen		3 260	3 410		3 130	
Adult bird		3 205	3 360		3 090	2 860
Apparent crude protein digestibility — ACPD (%)					82	
Available phosphorus (%)		**0.07**	**0.08**	**0.15**	**0.18**	**0.08**

Table 93 (cont.)

Minor cereals	Code:	13	14	15	16	17
Characteristics (% fresh weight)		Barley — naked	Rice	Buck-wheat	Rye — winter	Triticale — French
Dry matter		86	87	87	86	86
Gross energy (kcal/kg)		3 800	3 700	3 930	3 720	3 770
Nitrogen-free extract		69.2	63.7	61.0	70.8	68.3
Starch–acid hydrolysis		50.0	41.8	49.5	52.5	53.5
Starch–enzymic hydrolysis		51.0				
Sugars		2.3	4.0	1.5	3.0	4.7
Ether extract		2.0	2.1	2.4	1.6	1.6
Crude fibre		2.2	8.6	10.1	2.4	2.7
Acid detergent fibre (ADF)		1.9	10.0	14.1		4.4
Neutral detergent figre (NDF)		10.6	17.0	17.8		11.5
Crude protein		**10.8**	**8.0**	**11.4**	**9.5**	**11.6**
Amino acids:						
Lysine		0.39	0.28	0.58	0.36	0.39
Methionine		0.18	0.16	0.15	0.17	0.20
Methionine + cystine		0.45	0.35	0.46	0.38	0.46
Tryptophan		0.12	0.09	0.16	0.10	0.11
Threonine		0.37	0.28	0.39	0.31	0.35
Glycine + serine		0.87	0.76	1.08	0.83	1.00
Leucine		0.74	0.58	0.65	0.59	0.73
Isoleucine		0.41	0.33	0.36	0.36	0.43
Valine		0.54	0.47	0.51	0.47	0.52
Histidine		0.23	0.17	0.25	0.21	0.29
Arginine		0.53	0.57	0.91	0.49	0.64
Phenylalanine + tyrosine		0.93	0.72	0.67	0.67	0.88
Ash		**1.80**	**4.53**	**2.11**	**1.72**	**1.80**
Calcium		0.04	0.05	0.08	0.06	0.04
Total phosphorus		0.30	0.26	0.30	0.34	0.40
Sodium		0.05	0.03	0.01	0.02	0.03
Potassium		0.45	0.30	0.40	0.45	
Chloride		0.10	0.06	0.05	0.02	
Magnesium		0.12	0.12	0.20	0.11	
Pig						
Digestible energy — DE (kcal/kg)		3 300	2 750	2 600	3 150	3 200
Metabolizable energy — ME (kcal/kg)		3 205	2 685	2 520	3 070	3 105
Apparent crude protein digestibility — ACPD (%)		81	66	68	71	79
Rabbit						
Digestible energy — DE (kcal/kg)		3 220		2 900	3 220	3 070
Indigestible crude fibre — IDCF (%)		1.9		9.0	1.4	1.0
Apparent crude protein digestibility — ACPD (%)		80		72	70	83
Poultry						
Metabolizable energy — ME (kcal/kg):						
Broiler						
Pullet and laying hen		2 950	2 780		2 785	3 025
Adult bird		2 905	2 750	2 480	2 750	2 975
Apparent crude protein digestibility — ACPD (%)						
Available phosphorus (%)		**0.12**	**0.10**	**0.10**	**0.17**	**0.22**

Table 93 (cont.)

Cereal by-products	Code:	18	19	20	21	22
Characteristics (% fresh weight)		Hard wheat			Soft wheat	
		Middlings	Shorts	Bran	Middlings	Germ
Dry matter		88	88	88	88	88
Gross energy (kcal/kg)		3 990	3 950	4 000	3 970	4 220
Nitrogen-free extract		65.2	54.6	53.3	67.0	45.8
Starch–acid hydrolysis		48.2	40.0	19.4	58.3	17.6
Starch–enzymic hydrolysis						
Sugars		3.5	4.0	6.5	4.0	13.0
Ether extract		3.5	4.8	4.7	2.7	8.3
Crude fibre		1.5	8.2	10.0	1.4	2.8
Acid detergent fibre (ADF)		1.9	9.9	12.1	1.7	3.2
Neutral detergent figre (NDF)		6.2	32.8	36.9	6.2	10.7
Crude protein		**15.4**	**16.4**	**15.6**	**14.9**	**26.7**
Amino acids:						
Lysine		0.47	0.74	0.65	0.50	1.63
Methionine		0.20	0.24	0.25	0.21	0.44
Methionine + cystine		0.52	0.56	0.62	0.46	0.90
Tryptophan		0.20	0.20	0.19	0.20	0.30
Threonine		0.43	0.52	0.54	0.41	0.93
Glycine + serine		1.30	1.45	1.42	1.29	2.39
Leucine		1.05	1.06	0.99	1.00	1.62
Isoleucine		0.54	0.56	0.53	0.53	0.93
Valine		0.70	0.77	0.73	0.68	1.36
Histidine		0.32	0.39	0.38	0.32	0.59
Arginine		0.62	1.03	1.04	0.73	1.83
Phenylalanine + tyrosine		1.20	1.08	1.10	1.10	1.70
Ash		**2.40**	**4.00**	**4.40**	**2.00**	**4.40**
Calcium		0.05	0.13	0.15	0.07	0.07
Total phosphorus		0.55	0.90	0.93	0.45	0.95
Sodium		0.05		0.01	0.05	0.02
Potassium		0.60		1.10	0.60	0.55
Chloride		0.05		0.06	0.06	0.07
Magnesium		0.20		0.35	0.16	0.27
Pig						
Digestible energy — DE (kcal/kg)		3 500	3 000	2 500	3 550	3 700
Metabolizable energy — ME (kcal/kg)		3 370	2 875	2 405	3 420	3 495
Apparent crude protein digestibility — ACPD (%)		90	82	65	90	90
Rabbit						
Digestible energy — DE (kcal/kg)					3 300	
Indigestible crude fibre — IDCF (%)					0.1	
Apparent crude protein digestibility — ACPD (%)					83	
Poultry						
Metabolizable energy — ME (kcal/kg):						
Broiler						
Pullet and laying hen		3 140	2 060	1 460	3 200	2 915
Adult bird		3 080	2 000	1 400	3 150	2 810
Apparent crude protein digestibility — ACPD (%)						
Available phosphorus (%)		**0.22**	**0.36**	**0.37**	**0.18**	**0.31**

Table 93 (cont.)

Cereal by-products	Code:	23	24	25	26	27
Characteristics (% fresh weight)		Soft wheat				Maize DDGS*
		White shorts	Red shorts	Fine bran	Coarse bran	
Dry matter		87	87	87	87	91
Gross energy (kcal/kg)		3 950	3 970	3 950	3 940	4 320
Nitrogen-free extract		60.8	56.1	52.8	51.9	45.4
Starch–acid hydrolysis		33.6	23.0	16.5	16.5	
Starch–enzymic hydrolysis						
Sugars		5.0	5.5	4.7	4.7	
Ether extract		3.8	4.4	4.0	4.0	7.9
Crude fibre		4.1	7.0	9.6	10.6	6.2
Acid detergent fibre (ADF)		5.0	8.8	12.0	13.9	
Neutral detergent figre (NDF)		18.0	30.8	40.0	46.1	
Crude protein		**15.5**	**15.1**	**15.0**	**14.7**	**26.9**
Amino acids:						
Lysine		0.58	0.69	0.56	0.55	0.70
Methionine		0.22	0.23	0.20	0.20	0.54
Methionine + cystine		0.52	0.52	0.50	0.49	0.91
Tryptophan		0.20	0.20	0.24	0.24	0.21
Threonine		0.50	0.52	0.54	0.53	0.97
Glycine + serine		1.43	1.51	1.50	1.47	2.29
Leucine		1.00	0.96	0.95	0.93	2.25
Isoleucine		0.55	0.51	0.52	0.51	1.26
Valine		0.80	0.79	0.73	0.72	1.47
Histidine		0.38	0.39	0.39	0.38	0.62
Arginine		1.01	1.04	1.05	1.03	1.00
Phenylalanine + tyrosine		1.13	1.07	1.02	1.00	2.18
Ash		**2.80**	**4.40**	**5.60**	**5.80**	**4.60**
Calcium		0.10	0.11	0.13	0.14	0.16
Total phosphorus		0.67	0.85	1.20	1.30	0.70
Sodium		0.05	0.05	0.04	0.01	0.05
Potassium		0.70	0.95	1.27	1.20	0.90
Chloride		0.04	0.04	0.07	0.06	0.18
Magnesium		0.20	0.32	0.50	0.40	0.28
Pig						
Digestible energy — DE (kcal/kg)		3 100	2 800	2 450	2 300	2 910
Metabolizable energy — ME (kcal/kg)		2 980	2 690	2 355	2 210	2 740
Apparent crude protein digestibility — ACPD (%)		83	80	67	65	75
Rabbit						
Digestible energy — DE (kcal/kg)				2 200	2 180	
Indigestible crude fibre—IDCF (%)				6.8	7.4	
Apparent crude protein digestibility — ACPD (%)				83	76	
Poultry						
Metabolizable energy — ME (kcal/kg):						
Broiler						
Pullet and laying hen		2 775	2 160	1 500	1 500	
Adult bird		2 720	2 100	1 440	1 440	2 375
Apparent crude protein digestibility — ACPD (%)						
Available phosphorus (%)		**0.27**	**0.35**	**0.60**	**0.60**	**0.60**

*Distilled dried grains and solubles.

Table 93 (cont.)

Cereal by-products	Code:	28	29	30	31	32
Characteristics (% fresh weight)				Maize		
		Germ	Gluten feed	Gluten meal 40	Gluten meal 60	Cobs
Dry matter		89	90	90	90	90
Gross energy (kcal/kg)		4 830	4 050	4 540	4 820	4 000
Nitrogen-free extract		40.1	52.3	37.7	21.7	53.1
Starch–acid hydrolysis						
Starch–enzymic hydrolysis						
Sugars						
Ether extract		20.7	3.0	2.8	2.7	0.8
Crude fibre		5.9	8.3	4.0	1.7	31.6
Acid detergent fibre (ADF)			10.0	5.0	2.1	42.1
Neutral detergent figre (NDF)			30.1	14.4	6.1	77.0
Crude protein		**15.4**	**21.0**	**42.7**	**61.9**	**2.9**
Amino acids:						
Lysine		0.71	0.69	0.77	1.00	0.11
Methionine		0.28	0.39	1.02	1.63	0.04
Methionine + cystine		0.60	0.97	1.65	2.92	0.10
Tryptophan		0.15	0.16	0.21	0.31	
Threonine		0.60	0.83	1.48	2.14	0.13
Glycine + serine		1.53	1.91	3.77	5.04	0.30
Leucine		1.31	2.10	7.20	10.40	0.23
Isoleucine		0.54	0.68	2.12	2.63	0.12
Valine		0.92	1.05	2.22	3.09	0.16
Histidine		0.46	0.72	1.01	1.27	0.06
Arginine		1.06	0.83	1.41	1.98	0.13
Phenylalanine + tyrosine		1.22	1.43	4.69	7.86	0.22
Ash		**6.92**	**5.44**	**2.79**	**1.93**	**1.65**
Calcium			0.28	0.15	0.02	0.11
Total phosphorus			0.70	0.43	0.37	0.06
Sodium			0.10	0.08	0.02	0.02
Potassium			0.60	0.03	0.03	0.70
Chloride				0.10	0.05	0.06
Magnesium			0.35	0.08	0.06	0.06
Pig						
Digestible energy — DE (kcal/kg)		3 770	2 600	3 700	4 250	1 530
Metabolizable energy — ME (kcal/kg)		3 645	2 455	3 395	3 790	1 520
Apparent crude protein digestibility — ACPD (%)		80	80	90	96	33
Rabbit						
Digestible energy — DE (kcal/kg)						
Indigestible crude fibre — IDCF (%)						
Apparent crude protein digestibility — ACPD (%)						
Poultry						
Metabolizable energy — ME (kcal/kg):						
Broiler					3 920	
Pullet and laying hen					3 845	
Adult bird		2 720			3 610	
Apparent crude protein digestibility — ACPD (%)						
Available phosphorus (%)			0.23	0.13	0.12	0.02

Table 93 (cont.)

Cereal by-products Characteristics (% fresh weight)	Code:	33 Bran	34 Maize Germ oilmeal*	35 Germ oilmeal**	36 Barley brewer's grains
Dry matter		89	91	91	91
Gross energy (kcal/kg)		4 080	4 450	4 120	4 500
Nitrogen-free extract		60.9	49.7	56.3	38.9
Starch–acid hydrolysis			11.0		8.3
Starch–enzymic hydrolysis					
Sugars			1.0		4.8
Ether extract		6.3	7.5	1.8	7.6
Crude fibre		9.0	10.6	10.5	15.3
Acid detergent fibre (ADF)		11.5	13.9	12.9	19.4
Neutral detergent figre (NDF)		36.5	47.0	43.6	47.0
Crude protein		**10.1**	**20.8**	**19.5**	**25.2**
Amino acids:					
Lysine		0.27	0.96	0.90	0.70
Methionine		0.15	0.37	0.35	0.35
Methionine + cystine		0.36	0.81	0.76	0.61
Tryptophan			0.21	0.20	0.25
Threonine			0.81	0.76	0.74
Glycine + serine			2.06	1.93	1.80
Leucine			1.77	1.66	1.93
Isoleucine			0.73	0.68	1.19
Valine			1.25	1.17	1.17
Histidine			0.62	0.59	0.43
Arginine			1.44	1.35	1.00
Phenylalanine + tyrosine			1.64	1.54	1.75
Ash		**2.69**	**2.36**	**2.94**	**4.07**
Calcium		0.03	0.06	0.09	0.28
Total phosphorus		0.23	0.45	0.50	0.50
Sodium			0.03	0.03	0.26
Potassium		0.65	0.18	0.25	0.09
Chloride		0.05	0.06	0.08	0.12
Magnesium		0.21	0.22	0.15	0.15
Pig					
Digestible energy — DE (kcal/kg)		2 500	3 000	2 600	2 050
Metabolizable energy — ME (kcal/kg)		2 420	2 860	2 470	1 900
Apparent crude protein digestibility — ACPD (%)		75	75	76	73
Rabbit					
Digestible energy — DE (kcal/kg)					
Indigestible crude fibre — IDCF (%)					
Apparent crude protein digestibility — ACPD (%)					
Poultry					
Metabolizable energy — ME (kcal/kg):					
Broiler					
Pullet and laying hen					2 435
Adult bird			1 800		2 340
Apparent crude protein digestibility — ACPD (%)					
Available phosphorus (%)		0.07	0.15	0.17	**0.40**

*Expeller. **Solvent.

Table 93 (cont.)

Cereal by-products	Code:	37	38	39	40
Characteristics (% fresh weight)		Barley rootlets	Rice		
			Broken	Shorts	Bran
Dry matter		91	89	89	90
Gross energy (kcal/kg)		4 070	3 840	4 300	4 170
Nitrogen-free extract		43.3	77.3	45.8	41.1
Starch–acid hydrolysis		4.5	69.0	20.4	
Starch–enzymic hydrolysis					
Sugars		13.8	1.5	5.9	
Ether extract		1.8	1.2	14.6	13.8
Crude fibre		13.1	1.3	6.7	11.6
Acid detergent fibre (ADF)		14.3			
Neutral detergent figre (NDF)		34.5			
Crude protein		**26.6**	**7.8**	**13.3**	**12.8**
Amino acids:					
Lysine		1.25	0.30	0.55	0.56
Methionine		0.38	0.20	0.27	0.22
Methionine + cystine		0.69	0.33	0.56	0.42
Tryptophan		0.32	0.12	0.13	0.13
Threonine		0.95	0.30	0.48	0.44
Glycine + serine		2.07	0.75	1.25	1.22
Leucine		1.52	0.67	0.88	0.90
Isoleucine		0.95	0.37	0.44	0.51
Valine		1.32	0.52	0.76	0.77
Histidine		0.48	0.18	0.35	0.34
Arginine		1.22	0.52	1.04	0.97
Phenylalanine + tyrosine		1.42	0.79	1.05	1.19
Ash		**6.15**	**1.41**	**8.60**	**10.70**
Calcium		0.25	0.04	0.04	0.07
Total phosphorus		0.70	0.15	1.30	1.40
Sodium		0.06		0.11	0.05
Potassium		0.20	0.15	1.15	1.50
Chloride		0.50		0.13	0.06
Magnesium		0.15	0.07	0.60	0.85
Pig					
Digestible energy — DE (kcal/kg)		2 850		3 320	2 800
Metabolizable energy — ME (kcal/kg)		2 680		3 215	2 705
Apparent crude protein digestibility — ACPD (%)		75		77	
Rabbit					
Digestible energy — DE (kcal/kg)					
Indigestible crude fibre — IDCF (%)					
Apparent crude protein digestibility — ACPD (%)					
Poultry					
Metabolizable energy — ME (kcal/kg):					
Broiler			3 220		
Pullet and laying hen			3 220		
Adult bird			3 180	2 980	
Apparent crude protein digestibility — ACPD (%)					
Available phosphorus (%)		0.23	**0.02**	**0.19**	**0.14**

Table 93 (*cont.*)

Ensiled cereals	Code:	41	42	43	44	45
Characteristics (% fresh weight)				Maize		
		Ears + stalk tops	Ears with husks	Ears without husks	Grains	Whole plant
Dry matter		35	51	52	62	27
Gross energy (kcal/kg)		1 560	2 300	2 330	2 790	1 150
Nitrogen-free extract		23.1	37.8	39.6	50.6	16.2
Starch–acid hydrolysis						
Starch–enzymic hydrolysis						
Sugars						
Ether extract		1.0	1.8	2.2	2.8	0.7
Crude fibre		6.1	5.6	3.9	1.4	5.9
Acid detergent fibre (ADF)			6.0		2.3	5.4
Neutral detergent figre (NDF)			16.2		6.8	11.6
Crude protein		**3.3**	**4.6**	**5.5**	**6.2**	**2.4**
Amino acids:						
Lysine		0.08	0.12	0.13	0.19	0.05
Methionine				0.08	0.12	0.05
Methionine + cystine		0.13		0.18	0.29	0.09
Tryptophan		0.03			0.05	0.03
Threonine		0.12		0.19	0.23	0.09
Glycine + serine						
Leucine		0.36		0.98	0.80	0.22
Isoleucine		0.13		0.18	0.25	0.09
Valine		0.17		0.26	0.32	0.12
Histidine		0.07		0.17	0.18	0.04
Arginine		0.12		0.21	0.29	0.07
Phenylalanine + tyrosine		0.29		0.40	0.60	0.20
Ash		**1.40**	**1.17**	**0.91**	**1.05**	**1.82**
Calcium		0.04	0.01	0.01	0.01	0.08
Total phosphorus		0.10	0.17	0.18	0.20	0.06
Sodium			0.005	0.005	0.01	
Potassium		0.30	0.26	0.27	0.25	0.34
Chloride					0.04	0.04
Magnesium		0.04	0.05	0.08	0.08	0.04
Pig						
Digestible energy — DE (kcal/kg)		870	1 630	1 770	2 460	620
Metabolizable energy — ME (kcal/kg)		845	1 590	1 720	2 400	610
Apparent crude protein digestibility — ACPD (%)					81	
Rabbit						
Digestible energy — DE (kcal/kg)						
Indigestible crude fibre — IDCF (%)						
Apparent crude protein digestibility — ACPD (%)						
Poultry						
Metabolizable energy — ME (kcal/kg):						
Broiler						
Pullet and laying hen						
Adult bird						
Apparent crude protein digestibility — ACPD (%)						
Available phosphorus (%)						

Table 93 (*cont.*)

Carbohydrate raw materials	Code:	46	47	48	49	50
Characteristics		Maize	Bananas			Carob
(% fresh weight)		starch				fruit
			Ensiled unripe	Whole unripe	Whole ripe	
Dry matter		88	29	21	22	86
Gross energy (kcal/kg)		3 670	1 200	870	900	3 620
Nitrogen-free extract		86.2		17.9	19.2	67.9
Starch–acid hydrolysis			20.6	15.4	1.5	
Starch–enzymic hydrolysis						
Sugars				0.4	14.8	30.3
Ether extract		0.9		0.3	0.2	2.4
Crude fibre		0.2	1.5	0.6	0.8	7.8
Acid detergent fibre (ADF)			2.4	1.1	1.8	38.4
Neutral detergent figre (NDF)			4.2	1.6	2.3	38.5
Crude protein		**0.6**	**1.5**	**1.2**	**1.3**	**5.0**
Amino acids:						
Lysine				0.05	0.06	
Methionine				0.01	0.01	
Methionine + cystine				0.03	0.03	
Tryptophan						
Threonine				0.04	0.05	
Glycine + serine						
Leucine				0.06	0.08	
Isoleucine				0.03	0.04	
Valine				0.04	0.06	
Histidine				0.07	0.11	
Arginine				0.06	0.07	
Phenylalanine + tyrosine				0.06	0.09	
Ash		0.12	1.10	1.00	1.10	3.43
Calcium					0.01	0.65
Total phosphorus					0.03	0.10
Sodium					0.01	0.03
Potassium					0.60	0.55
Chloride					0.18	0.15
Magnesium					0.40	0.04
Pig						
Digestible energy — DE (kcal/kg)		3 600		670	750	1 990
Metabolizable energy — ME (kcal/kg)		3 600		670	750	1 950
Apparent crude protein digestibility — ACPD (%)					30	
Rabbit						
Digestible energy — DE (kcal/kg)						2 390
Indigestible crude fibre — IDCF (%)						7.0
Apparent crude protein digestibility — ACPD (%)						61
Poultry						
Metabolizable energy — ME (kcal/kg):						
Broiler						
Pullet and laying hen						
Adult bird		3 490				
Apparent crude protein digestibility — ACPD (%)						
Available phosphorus (%)						0.03

Table 93 (cont.)

Carbohydrate raw materials	Code:	51	52	53	54	55
Characteristics (% fresh weight)		Carob germ	Chestnut		Bread-fruit	Acorns — dehulled
			Dehulled	Whole		
Dry matter		90	90	87	26	89
Gross energy (kcal/kg)		4 470	4 070	3 790		4 000
Nitrogen-free extract		30.8	74.3	70.4	17.6	72.3
Starch–acid hydrolysis					16.5	57.5
Starch–enzymic hydrolysis						
Sugars		9.7			0.6	
Ether extract		4.5	4.1	2.6	1.5	4.3
Crude fibre		3.3	2.4	5.0	4.6	3.8
Acid detergent fibre (ADF)		7.0				
Neutral detergent figre (NDF)		16.0				
Crude protein		**45.4**	**6.8**	**6.2**	**1.2**	**6.1**
Amino acids:						
Lysine		2.44	0.42			0.28
Methionine		0.35	0.11			0.10
Methionine + cystine		0.88	0.18			0.23
Tryptophan		0.05	0.09			0.06
Threonine		1.28	0.37			0.19
Glycine + serine		4.30				0.39
Leucine		2.66				0.37
Isoleucine		1.49				0.25
Valine		1.87	0.44			0.30
Histidine		1.07	0.21			0.13
Arginine		5.59	0.53			0.35
Phenylalanine + tyrosine		2.57				0.34
Ash		**6.00**	**2.40**	**2.84**	**1.07**	**2.50**
Calcium		0.60		0.23	0.03	
Total phosphorus		0.95		0.27	0.03	
Sodium		0.03		0.02		
Potassium		1.50		1.35	0.04	
Chloride		0.06		0.10		
Magnesium		0.30		0.02		
Pig						
Digestible energy — DE (kcal/kg)		4 180	3 300	3 050	670	2 900
Metabolizable energy — ME (kcal/kg)		3 850	3 240	3 000	670	2 845
Apparent crude protein digestibility — ACPD (%)		91	60			
Rabbit						
Digestible energy — DE (kcal/kg)						
Indigestible crude fibre — IDCF (%)						
Apparent crude protein digestibility — ACPD (%)						
Poultry						
Metabolizable energy — ME (kcal/kg):						
Broiler						
Pullet and laying hen						
Adult bird						
Apparent crude protein digestibility — ACPD (%)						
Available phosphorus (%)		0.31		0.09		

Table 93 (cont.)

Carbohydrate raw materials	Code: 56	57	58
Characteristics (% fresh weight)	Whole acorns	Dasheen	Sugarbeet molasses*
Dry matter	88	27	77
Gross energy (kcal/kg)	3 950	1 290	2 990
Nitrogen-free extract	64.1	23.7	60.0
Starch–acid hydrolysis		15.9	
Starch–enzymic hydrolysis			
Sugars		1.9	48.5
Ether extract	3.3	0.3	0.3
Crude fibre	12.0	1.3	0.3
Acid detergent fibre (ADF)			
Neutral detergent figre (NDF)			
Crude protein	**5.9**	**1.7**	**7.7**
Amino acids:			
Lysine			0.04
Methionine			0.05
Methionine + cystine			0.10
Tryptophan			0.10
Threonine			0.06
Glycine + serine			0.36
Leucine			0.22
Isoleucine			0.23
Valine			0.17
Histidine			0.00
Arginine			0.02
Phenylalanine + tyrosine			0.34
Ash	**2.70**	**1.60**	**8.93**
Calcium	0.10	0.04	0.25*
Total phosphorus	0.13	0.04	0.02
Sodium	0.01		1.00*
Potassium	1.00	0.06	4.00*
Chloride		0.07	1.30
Magnesium	0.06		0.03–0.25*
Pig			
Digestible energy — DE (kcal/kg)		1 050	2 600
Metabolizable energy — ME (kcal/kg)		1 050	2 530
Apparent crude protein digestibility — ACPD (%)			85
Rabbit			
Digestible energy — DE (kcal/kg)			2 600
Indigestible crude fibre — IDCF (%)			0
Apparent crude protein digestibility — ACPD (%)			51
Poultry			
Metabolizable energy — ME (kcal/kg):			
Broiler			
Pullet and laying hen			
Adult bird			2 250
Apparent crude protein digestibility — ACPD (%)			
Available phosphorus (%)	0.04	0.01	**0.01**

*With molasses from "Quentin" variety, the level of magnesium is higher (0.75%) and the levels of calcium, sodium and potassium lower (0.05, 0.4 and 1.0% respectively).

Table 93 (*cont.*)

Carbohydrate raw materials	Code:	59	60	61
Characteristics *(% fresh weight)*		Sugar cane molasses	Apples	Sugar
Dry matter		75	22	99
Gross energy (kcal/kg)		2 850	350	4 020
Nitrogen-free extract		63.2	10.5	99.0
Starch–acid hydrolysis			0.1	
Starch–enzymic hydrolysis				
Sugars		46.0	9.1	99.0
Ether extract		0	0.40	
Crude fibre		0	1.5	
Acid detergent fibre (ADF)				
Neutral detergent figre (NDF)				
Crude protein		3.4	0.5	
Amino acids:				
Lysine		0.02		
Methionine		0.02		
Methionine + cystine		0.04		
Tryptophan				
Threonine		0.05		
Glycine + serine		0.12		
Leucine		0.04		
Isoleucine		0.04		
Valine		0.08		
Histidine		0.01		
Arginine		0.02		
Phenylalanine + tyrosine		0.05		
Ash		8.42	0.41	
Calcium		0.70	0.01	
Total phosphorus		0.08	0.01	
Sodium		0.13	0.01	
Potassium		2.50	0.13	
Chloride		2.00	0.01	
Magnesium		0.35	0.02	
Pig				
Digestible energy — DE (kcal/kg)		2 530	300	3 920
Metabolizable energy — ME (kcal/kg)		2 530	300	3 920
Apparent crude protein digestibility — ACPD (%)				
Rabbit				
Digestible energy — DE (kcal/kg)		2 550		
Indigestible crude fibre — IDCF (%)		0		
Apparent crude protein digestibility — ACPD (%)		51		
Poultry				
Metabolizable energy — ME (kcal/kg):				
Broiler				3 900
Pullet and laying hen				3 900
Adult bird		2 230		3 900
Apparent crude protein digestibility — ACPD (%)				
Available phosphorus (%)		0.04		

Table 93 (*cont.*)

Fats/oils	Code:	62	63	64	65	66
Characteristics (% fresh weight)		Animal fat	Poultry fat	Vegetable oil	Lard	Tallow
Dry matter		99	99	99	99	99
Gross energy (kcal/kg)		9 370	9 400	9 450	9 440	9 410
Nitrogen-free extract						
Starch–acid hydrolysis						
Starch–enzymic hydrolysis						
Sugars						
Ether extract		98.4	98.5	98.7	98.5	98.3
Crude fibre						
Acid detergent fibre (ADF)						
Neutral detergent figre (NDF)						
Crude protein						
Amino acids:						
Lysine						
Methionine						
Methionine + cystine						
Tryptophan						
Threonine						
Glycine + serine						
Leucine						
Isoleucine						
Valine						
Histidine						
Arginine						
Phenylalanine + tyrosine						
Ash						
Calcium						
Total phosphorus						
Sodium						
Potassium						
Chloride						
Magnesium						
Pig						
Digestible energy — DE (kcal/kg)		7 900	8 200	8 500	7 900	7 500
Metabolizable energy — ME (kcal/kg)		7 900	8 200	8 500	7 900	7 500
Apparent crude protein digestibility — ACPD (%)						
Rabbit						
Digestible energy — DE (kcal/kg)						
Indigestible crude fibre — IDCF (%)						
Apparent crude protein digestibility — ACPD (%)						
Poultry						
Metabolizable energy — ME (kcal/kg):						
Broiler		8 450	9 080	9 200	8 550	7 020
Pullet and laying hen		8 500	9 100	9 250	8 600	7 320
Adult bird		8 530	9 200	9 250	8 700	8 040
Apparent crude protein digestibility — ACPD (%)						
Available phosphorus (%)						

Table 93 (cont.)

Roots and tubers	Code:	67	68	69	70	71
Characteristics (% fresh weight)			Beet			Arrow-root
		Sugar	Fodder sugar	Fodder	Sugar pulp	
Dry matter		24	19	13	90	22
Gross energy (kcal/kg)		960	750	520	3 785	900
Nitrogen-free extract		19.3	14.9	9.7	56.8	20.0
Starch–acid hydrolysis		0.1			1.0	18.1
Starch–enzymic hydrolysis						
Sugars		15.5	7.1	7.5	7.0	1.3
Ether extract		0.1	0.1	0.2	1.0	0.1
Crude fibre		2.0	1.1	1.0	18.0	0.6
Acid detergent fibre (ADF)					21.0	
Neutral detergent figre (NDF)					37.0	
Crude protein		**1.5**	**1.7**	**1.4**	**8.8**	**0.8**
Amino acids:						
Lysine		0.08	0.06	0.06	0.54	0.03
Methionine		0.02	0.01	0.02	0.07	0.02
Methionine + cystine		0.04	0.02	0.04	0.13	0.03
Tryptophan		0.02	0.02	0.10	0.09	
Threonine		0.06	0.04	0.03	0.39	0.04
Glycine + serine		0.20	0.21	0.19	0.63	
Leucine		0.09	0.05	0.05	0.57	0.05
Isoleucine		0.05	0.04	0.05	0.30	0.03
Valine		0.06	0.05	0.05	0.45	0.04
Histidine		0.03	0.02	0.02	0.22	0.01
Arginine		0.05	0.04	0.04	0.31	0.03
Phenylalanine + tyrosine		0.09	0.07	0.07	0.67	0.06
Ash		**1.10**	**1.20**	**1.20**	**5.42**	**0.94**
Calcium		0.05	0.04	0.03	0.90	0.03
Total phosphorus		0.04	0.03	0.03	0.11	0.02
Sodium		0.05	0.06	0.08	0.16	
Potassium		0.23	0.40	0.33	0.20	0.40
Chloride		0.08	0.08	0.08	0.10	
Magnesium		0.03	0.02	0.02	0.30	
Pig						
Digestible energy — DE (kcal/kg)		820	650	430	2 300	660
Metabolizable energy — ME (kcal/kg)		820	650	430	2 250	660
Apparent crude protein digestibility — ACPD (%)		47	50	62	40	
Rabbit						
Digestible energy — DE (kcal/kg)				492	2 900	
Indigestible crude fibre — IDCF (%)				0	5.0	
Apparent crude protein digestibility — ACPD (%)				66	87	
Poultry						
Metabolizable energy — ME (kcal/kg):						
Broiler						
Pullet and laying hen						
Adult bird						
Apparent crude protein digestibility — ACPD (%)						
Available phosphorus (%)					0.04	

Table 93 (cont.)

Roots and tubers	Code:	72	73	74	75	76
Characteristics (% fresh weight)		Carrots	Chicory		Yam	Cassava—pelleted
			Leaves	Roots		
Dry matter		12	17	33	25	85
Gross energy (kcal/kg)		500	620	1 350	1 040	3 400
Nitrogen-free extract		9.0	9.5	24.4	21.0	71.9
Starch–acid hydrolysis					16.9	62.0
Starch–enzymic hydrolysis						
Sugars					0.5	2.5
Ether extract		0.2		1.0		0.7
Crude fibre		1.0	1.9	1.9	0.7	4.6
Acid detergent fibre (ADF)						5.0
Neutral detergent figre (NDF)					2.7	10.0
Crude protein		**1.1**	**2.3**	**2.6**	**2.5**	**2.6**
Amino acids:						
Lysine		0.05			0.12	0.09
Methionine		0.01			0.04	0.03
Methionine + cystine					0.08	0.06
Tryptophan		0.02				0.02
Threonine		0.01			0.11	0.07
Glycine + serine		0.04				0.17
Leucine		0.05			0.16	0.12
Isoleucine		0.04			0.09	0.07
Valine		0.04			0.11	0.09
Histidine		0.02			0.06	0.03
Arginine		0.04			0.21	0.12
Phenylalanine + tyrosine		0.06			0.21	0.12
Ash		**1.00**	**3.34**	**3.10**	**1.24**	**5.22**
Calcium		0.05	0.24			0.30
Total phosphorus		0.03	0.07		0.04	0.19
Sodium		0.05	0.01			0.04
Potassium		0.30	0.23		0.40	1.10
Chloride		0.06				0.10
Magnesium		0.02	0.04			0.13
Insoluble hydrochloride						2.50
Pig						
Digestible energy — DE (kcal/kg)		400			640	3 010
Metabolizable energy — ME (kcal/kg)		400			640	3 010
Apparent crude protein digestibility — ACPD (%)		62				
Rabbit						
Digestible energy — DE (kcal/kg)			230			2 850
Indigestible crude fibre — IDCF (%)			0.3			2.0
Apparent crude protein digestibility — ACPD (%)			84			70
Poultry						
Metabolizable energy — ME (kcal/kg):						
Broiler						2 900
Pullet and laying hen						2 890
Adult bird						2 850
Apparent crude protein digestibility — ACPD (%)						
Available phosphorus (%)						0.06

Table 93 (*cont.*)

Roots and tubers	Code:	77	78	79
Characteristics *(% fresh weight)*		Cassava roots	Turnips	Sweet potato
Dry matter		87	10	33
Gross energy (kcal/kg)		3 560	420	1 400
Nitrogen-free extract		78.0	6.5	29.2
Starch–acid hydrolysis		69.5		22.3
Starch–enzymic hydrolysis				
Sugars		3.5		3.0
Ether extract		0.7	0.3	0.4
Crude fibre		3.0	1.2	1.1
Acid detergent fibre (ADF)		4.0		1.4
Neutral detergent figre (NDF)		9.0		
Crude protein		**2.2**	**1.3**	**1.6**
Amino acids:				
Lysine		0.08		0.06
Methionine		0.03		0.02
Methionine + cystine		0.05		0.04
Tryptophan		0.02		0.03
Threonine		0.06		0.09
Glycine + serine		0.15		
Leucine		0.10		0.09
Isoleucine		0.06		0.06
Valine		0.08		0.08
Histidine		0.03		0.03
Arginine		0.10		0.09
Phenylalanine + tyrosine		0.10		0.13
Ash		**3.13**	**1.16**	**1.17**
Calcium		0.20	0.06	0.04
Total phosphorus		0.15	0.03	0.05
Sodium		0.03	0.06	0.02
Potassium		0.40	0.24	0.35
Chloride		0.06	0.07	0.02
Magnesium		0.08	0.007	0.05
Insoluble hydrochloride				0.04
Pig				
Digestible energy — DE (kcal/kg)		3 360	330	1 200
Metabolizable energy — ME (kcal/kg)		3 360	330	1 200
Apparent crude protein digestibility — ACPD (%)			60	36
Rabbit				
Digestible energy — DE (kcal/kg)				1 190
Indigestible crude fibre — IDCF (%)				0.1
Apparent crude protein digestibility — ACPD (%)				44
Poultry				
Metabolizable energy — ME (kcal/kg):				
Broiler				
Pullet and laying hen				
Adult bird				
Apparent crude protein digestibility — ACPD (%)				
Available phosphorus (%)		0.05		0.02

Table 93 (cont.)

Roots and tubers	Code:	80	81	82	83
Characteristics (% fresh weight)			Potatoes		
		Starch	Raw	Dried	Starch flour
Dry matter		83	24	89	86
Gross energy (kcal/kg)		3 380	990	3 700	3 470
Nitrogen-free extract		81.2	19.7	72.7	78.2
Starch–acid hydrolysis		80.5	17.7	62.8	
Starch–enzymic hydrolysis					
Sugars					
Ether extract		0.1	0.1	0.6	0.9
Crude fibre		0.6	0.7	2.7	2.2
Acid detergent fibre (ADF)					
Neutral detergent figre (NDF)					
Crude protein		**0.8**	**2.2**	**9.4**	**0.4**
Amino acids:					
Lysine			0.11	0.41	
Methionine			0.03	0.11	
Methionine + cystine			0.05	0.18	
Tryptophan			0.02	0.10	
Threonine			0.08	0.30	
Glycine + serine					
Leucine			0.13	0.48	
Isoleucine			0.10	0.37	
Valine			0.11	0.41	
Histidine			0.03	0.11	
Arginine			0.11	0.41	
Phenylalanine + tyrosine			0.15	0.56	
Ash		**0.25**	**1.29**	**4.10**	**4.42**
Calcium		0.02	0.02	0.08	
Total phosphorus		<0.01	0.05	0.15	
Sodium		<0.01	0.01	0.03	
Potassium		0.01	0.50	1.50	
Chloride			0.06	0.20	
Magnesium			0.03	0.10	
Pig					
Digestible energy — DE (kcal/kg)		3 180	860	3 230	
Metabolizable energy — ME (kcal/kg)		3 180	860	3 210	
Apparent crude protein digestibility — ACPD (%)			59	45	
Rabbit					
Digestible energy — DE (kcal/kg)				3 100	
Indigestible crude fibre — IDCF (%)				0.2	
Apparent crude protein digestibility — ACPD (%)				42	
Poultry					
Metabolizable energy — ME (kcal/kg):					
Broiler					
Pullet and laying hen					
Adult bird		1 825			
Apparent crude protein digestibility — ACPD (%)					
Available phosphorus (%)		0	0.02	0.05	

Table 93 (*cont.*)

Roots and tubers *Characteristics* *(% fresh weight)*	Code:	84	85	86	87	88
			Potatoes		Swedes	Jerusalem artichoke
		Flakes	Proteins	Pulp		
Dry matter		88	93	88	11	22
Gross energy (kcal/kg)		3 610	5 120	3 680	480	930
Nitrogen-free extract		74.2	10.0	63.9		17.3
Starch–acid hydrolysis						
Starch–enzymic hydrolysis						
Sugars						
Ether extract		0.3	3.2	0.5	0.3	0.3
Crude fibre		2.0	0.5	14.0		1.0
Acid detergent fibre (ADF)				18.6		
Neutral detergent figre (NDF)				36.7		
Crude protein		**7.7**	**76.5**	**5.2**	**1.3**	**1.8**
Amino acids:						
Lysine			5.40	0.26	0.46	
Methionine			1.50	0.06	0.01	
Methionine + cystine			2.79	0.10	0.02	
Tryptophan			0.92	0.03	0.01	
Threonine			3.86	0.16		
Glycine + serine				0.37		
Leucine			6.88	0.29		
Isoleucine			3.98	0.17		
Valine			4.82	0.27		
Histidine			1.53	0.10		
Arginine			3.98	0.17		
Phenylalanine + tyrosine			7.80	0.37		
Ash		**3.90**	**2.79**	**3.10**	**1.00**	**1.70**
Calcium		0.07	0.48	0.18		0.06
Total phosphorus		0.14	0.18	0.09		0.05
Sodium		0.01	0.03	0.05		0.10
Potassium		1.40	0.80	0.90		0.40
Chloride		0.20	0.20	0.04		0.06
Magnesium		0.10	0.04	0.06		0.02
Pig						
Digestible energy — DE (kcal/kg)		3 260	4 140	3 200	380	750
Metabolizable energy — ME (kcal/kg)		3 190	3 750	3 150	380	750
Apparent crude protein digestibility — ACPD (%)		67		50	64	46
Rabbit						
Digestible energy — DE (kcal/kg)		3 240				815
Indigestible crude fibre — IDCF (%)		0.7				0
Apparent crude protein digestibility — ACPD (%)		60				72
Poultry						
Metabolizable energy — ME (kcal/kg):						
Broiler		2 920				
Pullet and laying hen		2 915				
Adult bird		2 900	3 770			
Apparent crude protein digestibility — ACPD (%)						
Available phosphorus (%)		0.05	0.06	0.05		

Table 93 (cont.)

Industrial by-products	Code:	89	90	91	92
Characteristics (% fresh weight)		Dried citrus pulp	Coffee grounds	Apple pomace	Grape pips
Dry matter		90	94	90	90
Gross energy (kcal/kg)		3 880	5 870	4 090	4 900
Nitrogen-free extract		64.1	14.0	53.3	
Starch–acid hydrolysis					
Starch–enzymic hydrolysis					
Sugars				14.2	
Ether extract		3.0	20.5	4.4	11.2
Crude fibre		12.0	47.2	23.9	34.6
Acid detergent fibre (ADF)		14.0	58.1	42.0	
Neutral detergent figre (NDF)		21.0		47.6	
Crude protein		**6.0**	**11.6**	**5.7**	**9.9**
Amino acids:					
Lysine		0.25	0.20		0.36
Methionine		0.06	0.17		0.15
Methionine + cystine		0.10	0.26		0.36
Tryptophan		0.06			
Threonine		0.21			
Glycine + serine		0.57			
Leucine		0.36			
Isoleucine		0.22			
Valine		0.26			
Histidine		0.11			
Arginine		0.24			
Phenylalanine + tyrosine		0.43			
Ash		**5.45**	**1.20**	**2.70**	**2.90**
Calcium		2.10		0.17	0.70
Total phosphorus		0.12		0.14	0.19
Sodium		0.10		0.03	0.20
Potassium		0.55		0.58	1.50
Chloride		0.06		0.07	
Magnesium		0.17		0.08	
Pig					
Digestible energy — DE (kcal/kg)		2 900	1 040	1 450	
Metabolizable energy — ME (kcal/kg)		2 855	1 020	1 450	
Apparent crude protein digestibility — ACPD (%)		35	10	0	
Rabbit					
Digestible energy — DE (kcal/kg)		3 300	2 400	1 640	2 150
Indigestible crude fibre — IDCF (%)		5.1	22.8	19.4	31.8
Apparent crude protein digestibility — ACPD (%)		69	77	20	67
Poultry					
Metabolizable energy — ME (kcal/kg):					
Broiler					
Pullet and laying hen					
Adult bird					
Apparent crude protein digestibility — ACPD (%)					
Available phosphorus (%)		0.04			

Table 93 (*cont.*)

Industrial by-products	Code:	93	94	95	96
Characteristics (% fresh weight)		\multicolumn{2}{c}{Grape}	\multicolumn{2}{c}{Tomato}	
		Pulp	Grapeseed oil meal	Skins	Pips
Dry matter		88	89	96	90
Gross energy (kcal/kg)		4 000	4 100	4 580	4 400
Nitrogen-free extract		36.2	29.5	32.3	17.8
Starch–acid hydrolysis					
Starch–enzymic hydrolysis					
Sugars					
Ether extract		6.3	1.4	6.5	6.2
Crude fibre		24.7	44.8	43.4	25.5
Acid detergent fibre (ADF)			54.0		
Neutral detergent figre (NDF)			59.7		
Crude protein		12.3	10.0	9.4	34.0
Amino acids:					
Lysine			0.41		1.97
Methionine			0.17		
Methionine + cystine			0.38		0.60
Tryptophan					0.68
Threonine			0.20		0.98
Glycine + serine			0.91		
Leucine			0.61		2.07
Isoleucine			0.41		1.54
Valine			0.51		1.73
Histidine			0.21		0.88
Arginine			0.66		2.62
Phenylalanine + tyrosine			0.90		
Ash		8.44	3.40	4.40	6.45
Calcium		0.70	0.60		
Total phosphorus		0.35	0.12		
Sodium		0.01	0.01		
Potassium		1.60	0.60		
Chloride		0.15			
Magnesium		0.12	0.10		
Pig					
Digestible energy — DE (kcal/kg)		1 940			
Metabolizable energy — ME (kcal/kg)		1 870			
Apparent crude protein digestibility — ACPD (%)		58			
Rabbit					
Digestible energy — DE (kcal/kg)		1 000	820		
Indigestible crude fibre — IDCF (%)		24.0	34.9		
Apparent crude protein digestibility — ACPD (%)		5	59		
Poultry					
Metabolizable energy — ME (kcal/kg):					
Broiler					
Pullet and laying hen					1 650
Adult bird					1 520
Apparent crude protein digestibility — ACPD (%)					
Available phosphorus (%)		0.12	0.04		

Table 93 (*cont.*)

Industrial by-products	Code:	97	98	99	100
Characteristics (% *fresh weight*)		Tomato pulp	Vinasse	Vinasse waste	
				Sugar	Yeast
Dry matter		91	70	74	78
Gross energy (kcal/kg)		4 200	2 140	3 140	
Nitrogen-free extract		28.9	29.2	22.9	19.3
Starch–acid hydrolysis					
Starch–enzymic hydrolysis					
Sugars			27.0	6.4	3.6
Ether extract		0.60			
Crude fibre		37.0			
Acid detergent fibre (ADF)					
Neutral detergent figre (NDF)					
Crude protein		18.3	23.0	47.0	43.7
Amino acids:					
Lysine			0.11	0.28	0.26
Methionine			0.09	0.14	0.13
Methionine + cystine			0.17	0.28	0.26
Tryptophan					
Threonine			0.15	0.09	0.09
Glycine + serine			0.55	0.66	0.61
Leucine			0.17	0.47	0.44
Isoleucine			0.09	0.38	0.35
Valine			0.20	0.70	0.65
Histidine			0.21	0.09	0.09
Arginine			0.09	0.19	0.17
Phenylalanine + tyrosine			0.29	0.33	0.30
Ash		6.20	17.80	4.10	15.00
Calcium		0.40	0.46	0.03	
Total phosphorus		0.70	0.13	0.12	
Sodium			2.11	0.90	
Potassium		3.30	4.58	0.15	
Chloride				1.10	
Magnesium		0.18	0.02	0.30	
Pig					
Digestible energy — DE (kcal/kg)			1 100	2 400	2 100
Metabolizable energy — ME (kcal/kg)			1 015	2 165	1 880
Apparent crude protein digestibility — ACPD (%)			45	70	65
Rabbit					
Digestible energy — DE (kcal/kg)					
Indigestible crude fibre — IDCF (%)					
Apparent crude protein digestibility — ACPD (%)					
Poultry					
Metabolizable energy — ME (kcal/kg):					
Broiler					
Pullet and laying hen					
Adult bird			950	1 920	1 680
Apparent crude protein digestibility — ACPD (%)					
Available phosphorus (%)		0.23			

Table 93 (*cont.*)

Various vegetable products	Code: 101	102
Characteristics (% fresh weight)	Cocoa hulls	Rapeseed hulls
Dry matter	90	87
Gross energy (kcal/kg)	4 150	4 720
Nitrogen-free extract	46.4	16.2
Starch–acid hydrolysis		
Starch–enzymic hydrolysis		
Sugars		
Ether extract	4.5	18.1
Crude fibre	15.0	32.4
Acid detergent fibre (ADF)		48.9
Neutral detergent figre (NDF)		58.0
Crude protein	**16.5**	**17.1**
Amino acids:		
Lysine		1.11
Methionine		
Methionine + cystine		0.74
Tryptophan		
Threonine		
Glycine + serine		
Leucine		
Isoleucine		
Valine		
Histidine		
Arginine		
Phenylalanine + tyrosine		
Ash	**7.62**	**3.70**
Calcium	0.30	1.24
Total phosphorus	0.35	0.13
Sodium	0.08	
Potassium	2.50	
Chloride	0.15	
Magnesium	0.40	
Pig		
Digestible energy — DE (kcal/kg)		1 400
Metabolizable energy — ME (kcal/kg)		1 360
Apparent crude protein digestibility — ACPD (%)		20
Rabbit		
Digestible energy — DE (kcal/kg)	2 190	2 850
Indigestible crude fibre — IDCF (%)	14.0	27
Apparent crude protein digestibility — ACPD (%)	13	70
Poultry		
Metabolizable energy — ME (kcal/kg):		
Broiler		
Pullet and laying hen		
Adult bird		
Apparent crude protein digestibility — ACPD (%)		
Available phosphorus (%)	**0.04**	0.02

Table 93 (cont.)

Various vegetable products	Code: 103	104
Characteristics (% fresh weight)	Cabbages	Brussels sprouts
Dry matter	13	14
Gross energy (kcal/kg)	570	560
Nitrogen-free extract	6.7	6.7
Starch–acid hydrolysis		
Starch–enzymic hydrolysis		
Sugars		
Ether extract	0.3	0.3
Crude fibre	2.2	2.2
Acid detergent fibre (ADF)	1.9	
Neutral detergent figre (NDF)	2.3	
Crude protein	**2.3**	**2.4**
Amino acids:		
Lysine	0.08	
Methionine		
Methionine + cystine	0.05	
Tryptophan	0.01	
Threonine	0.06	
Glycine + serine	0.12	
Leucine	0.08	
Isoleucine	0.05	
Valine	0.08	
Histidine	0.04	
Arginine	0.09	
Phenylalanine + tyrosine	0.24	
Ash	**1.89**	**2.24**
Calcium	0.06	0.03
Total phosphorus	0.05	0.06
Sodium	0.01	0.005
Potassium	0.40	0.38
Chloride	0.03	0.03
Magnesium	0.02	0.02
Pig		
Digestible energy — DE (kcal/kg)		400
Metabolizable energy — ME (kcal/kg)		400
Apparent crude protein digestibility — ACPD (%)		69
Rabbit		
Digestible energy — DE (kcal/kg)		
Indigestible crude fibre — IDCF (%)		
Apparent crude protein digestibility — ACPD (%)		
Poultry		
Metabolizable energy — ME (kcal/kg):		
Broiler		
Pullet and laying hen		
Adult bird		
Apparent crude protein digestibility — ACPD (%)		
Available phosphorus (%)		

Table 93 (*cont.*)

Various vegetable products Code:	105	106	107	108
Characteristics *(% fresh weight)*	Cauli- flower	Marrow- stem kale	Cereal straw	Grass meal
Dry matter	13	12	88	91
Gross energy (kcal/kg)	510	490	3 950	3 860
Nitrogen-free extract	6.2	5.8		36.2
Starch–acid hydrolysis				
Starch–enzymic hydrolysis				
Sugars				
Ether extract	0.3	0.4		3.7
Crude fibre	2.0	1.6	42.0	21.0
Acid detergent fibre (ADF)			54.0	
Neutral detergent figre (NDF)			82.0	
Crude protein	**2.3**	**2.3**	4.5	**17.1**
Amino acids:				
Lysine			0.03	0.75
Methionine				0.27
Methionine + cystine				0.44
Tryptophan				
Threonine				0.68
Glycine + serine				1.64
Leucine				1.40
Isoleucine				0.68
Valine				1.15
Histidine				0.38
Arginine				0.61
Phenylalanine + tyrosine				1.49
Ash	**2.28**	**1.99**	**8.30**	**12.70**
Calcium	0.02		0.47	0.70
Total phosphorus	0.04		0.09	0.42
Sodium	0.01		0.11	0.06
Potassium	0.35		2.00	2.25
Chloride	0.03			0.08
Magnesium	0.01			0.20
Pig				
Digestible energy — DE (kcal/kg)	370	370		2 130
Metabolizable energy — ME (kcal/kg)	370	370		2 040
Apparent crude protein digestibility — ACPD (%)	69	68		58
Rabbit				
Digestible energy — DE (kcal/kg)			700	1 730
Indigestible crude fibre — IDCF (%)			39.5	14.1
Apparent crude protein digestibility — ACPD (%)			70	50
Poultry				
Metabolizable energy — ME (kcal/kg):				
Broiler				
Pullet and laying hen				
Adult bird				
Apparent crude protein digestibility — ACPD (%)				
Available phosphorus (%)				**0.40**

Table 93 (*cont.*)

Various vegetable products	Code:	109	110	111	112
Characteristics (% fresh weight)		Dehydrated Lucerne meal			
		21	17	15	12
Dry matter		90	90	90	90
Gross energy (kcal/kg)		3 980	3 960	3 940	3 925
Nitrogen-free extract		33.5	36.0	35.5	36.5
Starch–acid hydrolysis					
Starch–enzymic hydrolysis					
Sugars					
Ether extract		3.9	2.9	3.0	2.1
Crude fibre		21.2	25.0	27.0	30.4
Acid detergent fibre (ADF)		26.0	30.9	33.4	40.0
Neutral detergent figre (NDF)		33.0	40.2	43.5	56.0
Crude protein		**21.4**	**16.6**	**15.5**	**12.2**
Amino acids:					
Lysine		0.94	0.73	0.68	0.54
Methionine		0.31	0.24	0.22	0.18
Methionine + cystine		0.58	0.45	0.42	0.33
Tryptophan		0.36	0.28	0.26	0.21
Threonine		0.89	0.69	0.64	0.51
Glycine + serine		1.92	1.49	1.39	1.10
Leucine		1.51	1.17	1.09	0.86
Isoleucine		0.93	0.72	0.67	0.53
Valine		1.17	0.91	0.85	0.67
Histidine		0.43	0.33	0.31	0.24
Arginine		0.92	0.71	0.66	0.52
Phenylalanine + tyrosine		1.71	1.33	1.24	0.98
Ash		**10.00**	**9.45**	**9.00**	**8.84**
Calcium		2.00	1.50	1.40	1.40
Total phosphorus		0.26	0.25	0.25	0.25
Sodium		0.08	0.07	0.07	0.05
Potassium		2.40	1.90	1.90	1.70
Chloride			0.48		0.35
Magnesium		0.30	0.27	0.27	0.15
Pig					
Digestible energy — DE (kcal/kg)		2 220	1 860	1 680	1 370
Metabolizable energy — ME (kcal/kg)		2 080	1 750	1 590	1 310
Apparent crude protein digestibility — ACPD (%)					
Rabbit					
Digestible energy — DE (kcal/kg)		2 600	2 450	2 410	1 900
Indigestible crude fibre — IDCF (%)		14.0	20.5	22.1	25.9
Apparent crude protein digestibility — ACPD (%)		80	78	70	55
Poultry					
Metabolizable energy — ME (kcal/kg):					
Broiler					
Pullet and laying hen					
Adult bird		1 245	1 070	1 000	660
Apparent crude protein digestibility — ACPD (%)					
Available phosphorus (%)		**0.22**	**0.22**	**0.22**	**0.22**

Table 93 (cont.)

Various vegetable products	Code: 113	114
Characteristics (% fresh weight)	Lucerne protein concentrate	Soya-bean hulls
Dry matter	90	92
Gross energy (kcal/kg)	4 620	3 980
Nitrogen-free extract	18.2	37.6
Starch–acid hydrolysis	2.0	
Starch–enzymic hydrolysis		
Sugars	2.0	
Ether extract	9.0	2.0
Crude fibre	2.7	34.0
Acid detergent fibre (ADF)		43.0
Neutral detergent figre (NDF)		58.0
Crude protein	**46.0**	**12.7**
Amino acids:		
Lysine	3.08	0.70
Methionine	0.83	0.16
Methionine + cystine	1.51	0.35
Tryptophan	0.92	0.17
Threonine	2.36	0.50
Glycine + serine	4.55	
Leucine	4.14	0.99
Isoleucine	2.29	0.64
Valine	2.64	0.68
Histidine	1.14	0.34
Arginine	1.52	0.99
Phenylalanine + tyrosine	4.33	1.14
Ash	**14.10**	**5.69**
Calcium	4.00	0.40
Total phosphorus	0.80	0.17
Sodium	0.18	
Potassium	1.00	
Chloride	0.30	
Magnesium		
Pig		
Digestible energy — DE (kcal/kg)	3 360	1 890
Metabolizable energy — ME (kcal/kg)	3 040	1 840
Apparent crude protein digestibility — ACPD (%)	89	32
Rabbit		
Digestible energy — DE (kcal/kg)		1 820
Indigestible crude fibre — IDCF (%)		34.0
Apparent crude protein digestibility — ACPD (%)		42
Poultry		
Metabolizable energy — ME (kcal/kg):		
Broiler	2 635	
Pullet and laying hen	2 800	
Adult bird	2 620	
Apparent crude protein digestibility — ACPD (%)		
Available phosphorus (%)	**0.64**	**0.03**

Table 93 (cont.)

Vegetable proteins	Code:	115	116		117
Characteristics (% fresh weight)		Full fat rapeseed	Field beans		Toasted beans
			Mean	SE	
Dry matter		90	87		88
Gross energy (kcal/kg)		6 150	3 900	50	3 980
Nitrogen-free extract		16.2	48.5		57.0
Starch–acid hydrolysis			35.0		
Starch–enzymic hydrolysis			35.6		
Sugars					
Ether extract		43.2	1.3		1.7
Crude fibre		6.1	7.5	1.2	4.5
Acid detergent fibre (ADF)		7.6	8.3		4.5
Neutral detergent figre (NDF)		10.4	11.0		10.0
Crude protein		**19.5**	**26.4**	**1.8**	**22.6**
Amino acids:					
Lysine		1.09	1.66		1.53
Methionine		0.42	0.21		0.32
Methionine + cystine		0.96	0.53		0.59
Tryptophan		0.24	0.22		0.22
Threonine		0.87	0.93		1.04
Glycine + serine		1.84	2.39		2.64
Leucine		1.35	1.95		1.97
Isoleucine		0.80	1.18		1.23
Valine		1.02	1.25		1.36
Histidine		0.51	0.66		0.69
Arginine		1.21	2.48		1.40
Phenylalanine + tyrosine		1.36	2.02		2.15
Ash		**5.03**	**3.38**		**3.88**
Calcium		0.30	0.11		0.11
Total phosphorus		0.48	0.61		0.45
Sodium			0.01		0.02
Potassium			1.20		1.30
Chloride			0.07		0.05
Magnesium			0.18		0.12
Pig					
Digestible energy — DE (kcal/kg)		5 000	3 300		3 390
Metabolizable energy — ME (kcal/kg)		4 835	3 120		3 230
Apparent crude protein digestibility — ACPD (%)		82	81		81
Rabbit					
Digestible energy — DE (kcal/kg)			2 800		3 360
Indigestible crude fibre — IDCF (%)			5.0		3.1
Apparent crude protein digestibility — ACPD (%)			80		84
Poultry					
Metabolizable energy — ME (kcal/kg):					
Broiler		3 980			
Pullet and laying hen		3 960	2 750		
Adult bird		3 900	2 650		2 510
Apparent crude protein digestibility — ACPD (%)					
Available phosphorus (%)		**0.10**	**0.15**		0.15

Table 93 (cont.)

Vegetable proteins	Code:	118	119	120	
Characteristics (% fresh weight)		Lentils	Sweet white lupins	Smooth winter peas	
				Mean	SE
Dry matter		88	87	86	
Gross energy (kcal/kg)		4 040	4 350	3 800	30
Nitrogen-free extract		54.1	27.6	52.7	
Starch–acid hydrolysis		36.8		42.4	1.4
Starch–enzymic hydrolysis			0.3		
Sugars				2.1	
Ether extract		1.7	9.6	1.6	
Crude fibre		4.3	10.7	6.3	1.0
Acid detergent fibre (ADF)		4.7	15.4	8.7	
Neutral detergent figre (NDF)		8.7	18.8	12.0	
Crude protein		**24.7**	**35.7**	**22.0**	**1.2**
Amino acids:					
Lysine		1.68	1.68	1.60	
Methionine		0.25	0.28	0.25	
Methionine + cystine		0.47	0.80	0.59	
Tryptophan		0.28	0.28	0.20	
Threonine		1.09	1.29	0.87	
Glycine + serine		2.52	3.07	2.00	
Leucine		1.98	2.50	1.53	
Isoleucine		1.16	1.57	0.97	
Valine		1.32	1.50	1.01	
Histidine		0.68	0.82	0.52	
Arginine		1.90	3.75	2.12	
Phenylalanine + tyrosine		2.28	3.00	1.73	
Ash		**3.17**	**3.40**	**3.40**	
Calcium		0.10	0.18	0.08	
Total phosphorus		0.40	0.40	0.40	
Sodium		0.01	0.01	0.01	
Potassium		1.00	0.90	1.10	
Chloride		0.03	0.02	0.03	
Magnesium		0.13	0.15	0.12	
Pig					
Digestible energy — DE (kcal/kg)		3 610	3 630	3 390	
Metabolizable energy — ME (kcal/kg)		3 420	3 380	3 225	
Apparent crude protein digestibility — ACPD (%)		88	86	85	
Rabbit					
Digestible energy — DE (kcal/kg)			2 890	2 800	
Indigestible crude fibre — IDCF (%)			7.0	4.4	
Apparent crude protein digestibility — ACPD (%)			81	88	
Poultry					
Metabolizable energy — ME (kcal/kg):					
Broiler			2 250		
Pullet and laying hen			2 530	2 540	
Adult bird		2 585	2 415	2 460	120
Apparent crude protein digestibility — ACPD (%)					
Available phosphorus (%)		0.13	**0.08**	**0.15**	

Table 93 (*cont.*)

Vegetable proteins	Code:	121		122	123
Characteristics		*Smooth spring peas*		*Soya-beans*	
(% fresh weight)		Mean	SE	Full fat	Protein isolate
Dry matter		86		89	94
Gross energy (kcal/kg)		3 800	30	5 000	5 370
Nitrogen-free extract		53.5		22.2	
Starch–acid hydrolysis		43.9	1.4		
Starch–enzymic hydrolysis					
Sugars		2.1			
Ether extract		1.6		18.0	1.5
Crude fibre		5.5	1.0	6.0	1.5
Acid detergent fibre (ADF)		8.5		8.4	
Neutral detergent figre (NDF)		11.0		12.0	
Crude protein		**22.0**	**1.2**	**37.0**	**88.5**
Amino acids:					
Lysine		1.60		2.35	5.66
Methionine		0.25		0.52	1.06
Methionine + cystine		0.59		1.15	3.01
Tryptophan		0.20		0.48	1.19
Threonine		0.87		1.44	3.19
Glycine + serine		2.00		3.55	7.26
Leucine		1.53		2.85	7.08
Isoleucine		0.97		1.78	4.87
Valine		1.01		1.77	4.52
Histidine		0.52		0.91	2.39
Arginine		2.12		2.81	7.35
Phenylalanine + tyrosine		1.73		3.20	8.77
Ash		**3.40**		**4.45**	**3.00**
Calcium		0.08		0.25	0.19
Total phosphorus		0.45		0.57	0.62
Sodium		0.01		0.01	0.50
Potassium		1.10		1.50	0.38
Chloride		0.03		0.02	
Magnesium		0.12		0.29	0.14
Pig					
Digestible energy — DE (kcal/kg)		3 430		4 200	4 940
Metabolizable energy — ME (kcal/kg)		3 260		3 945	4 350
Apparent crude protein digestibility — ACPD (%)		88		85	
Rabbit					
Digestible energy — DE (kcal/kg)				4 400	
Indigestible crude fibre — IDCF (%)				3.6	
Apparent crude protein digestibility — ACPD (%)				88	
Poultry					
Metabolizable energy — ME (kcal/kg):					
Broiler				3 800	4 060
Pullet and laying hen		2 600		3 750	3 945
Adult bird		2 530	100	3 600	3 595
Apparent crude protein digestibility — ACPD (%)				80	85
Available phosphorus (%)		**0.14**		**0.11**	**0.10**

Table 93 (*cont.*)

Oil meals	Code:	124	125	126	127
Characteristics (% fresh weight)		Peanut meal	Rapeseed meal		
			Dehulled	Expeller	Solvent-extracted
Dry matter		91	89	89	89
Gross energy (kcal/kg)		4 360	4 120	4 490	4 150
Nitrogen-free extract		25.0	31.8	30.1	33.3
Starch–acid hydrolysis		6.9	5.3	4.8	5.2
Starch–enzymic hydrolysis					
Sugars			8.5	7.7	8.3
Ether extract		1.4	2.7	8.9	1.8
Crude fibre		10.0	5.8	11.1	11.7
Acid detergent fibre (ADF)		12.8	10.0	17.4	18.5
Neutral detergent figre (NDF)		17.0	13.8	24.4	25.5
Crude protein		**49.2**	**41.5**	**32.4**	**35.2**
Amino acids:					
Lysine		1.70	2.30	1.81	1.97
Methionine		0.49	1.00	0.70	0.76
Methionine + cystine		1.18	2.11	1.59	1.73
Tryptophan		0.49	0.51	0.40	0.43
Threonine		1.33	1.85	1.44	1.57
Glycine + serine		5.18	4.00	3.06	3.32
Leucine		3.09	2.92	2.24	2.43
Isoleucine		1.77	1.80	1.32	1.44
Valine		2.16	2.13	1.69	1.84
Histidine		1.14	1.20	0.85	0.92
Arginine		5.52	2.70	2.01	2.19
Phenylalanine + tyrosine		4.50	2.90	2.25	2.45
Ash		**5.40**	**7.16**	**6.51**	**7.00**
Calcium		0.16		0.75	0.75
Total phosphorus		0.60		1.05	1.10
Sodium		0.02		0.07	0.07
Potassium		1.15		1.25	1.25
Chloride		0.05		traces	traces
Magnesium		0.30		0.45	0.45
Pig					
Digestible energy — DE (kcal/kg)		3 640	3 520	3 330	2 850
Metabolizable energy — ME (kcal/kg)		3 295	3 230	3 115	2 630
Apparent crude protein digestibility — ACPD (%)		90	87	80	79
Rabbit					
Digestible energy — DE (kcal/kg)		3 700			2 920
Indigestible crude fibre — IDCF (%)		5.8			7.4
Apparent crude protein digestibility — ACPD (%)		94			79
Poultry					
Metabolizable energy — ME (kcal/kg):					
Broiler			1 630	1 865	1 350
Pullet and laying hen		2 825	1 860	2 100	1 580
Adult bird		2 650	1 780	2 015	1 490
Apparent crude protein digestibility — ACPD (%)		90	64	64	64
Available phosphorus (%)		0.06		**0.21**	**0.22**

Table 93 (cont.)

Oil meals	Code:	128	129	130
Characteristics (% fresh weight)		Copra meal	Cotton seed meal	Palm kernel meal
Dry matter		90	91	90
Gross energy (kcal/kg)		3 850	4 280	4 030
Nitrogen-free extract		43.9	29.1	50.9
Starch–acid hydrolysis				3.00
Starch–enzymic hydrolysis				
Sugars		9.0		1.0
Ether extract		1.6	1.4	1.7
Crude fibre		16.0	13.0	15.0
Acid detergent fibre (ADF)		30.0	21.1	31.7
Neutral detergent figre (NDF)		54.0	29.8	52.0
Crude protein		**21.5**	**41.0**	**18.5**
Amino acids:				
Lysine		0.66	1.72	0.66
Methionine		0.32	0.59	0.32
Methionine + cystine		0.63	1.24	0.70
Tryptophan		0.15	0.49	0.19
Threonine		0.67	1.39	0.61
Glycine + serine		1.81	3.53	1.75
Leucine		1.33	2.44	1.16
Isoleucine		0.76	1.39	0.69
Valine		1.18	1.97	1.00
Histidine		0.47	1.11	0.48
Arginine		2.62	4.33	2.56
Phenylalanine + tyrosine		1.50	3.30	1.15
Ash		**7.00**	**6.46**	**3.90**
Calcium		0.18	0.20	0.28
Total phosphorus		0.60	1.00	0.60
Sodium		0.04	0.05	0.02
Potassium		1.20	1.25	0.50
Chloride		0.10	0.04	0.13
Magnesium		0.30	0.50	0.35
Pig				
Digestible energy — DE (kcal/kg)		2 800	2 450	2 700
Metabolizable energy — ME (kcal/kg)		2 675	2 225	2 580
Apparent crude protein digestibility — ACPD (%)		63	70	70
Rabbit				
Digestible energy — DE (kcal/kg)		2 700	2 790	2 700
Indigestible crude fibre — IDCF (%)		7.0	9.0	9.0
Apparent crude protein digestibility — ACPD (%)		65	84	64
Poultry				
Metabolizable energy — ME (kcal/kg):				
Broiler				
Pullet and laying hen		1 330	1 945	1 240
Adult bird		1 250	1 800	1 170
Apparent crude protein digestibility — ACPD (%)				
Available phosphorus (%)		**0.09**	**0.10**	**0.09**

Table 93 (*cont.*)

Oil meals	Code:	131		132	
Characteristics *(% fresh weight)*		*Soya-bean meal 44*		*Soya-bean meal 48*	
		Mean	SE	Mean	SE
Dry matter		88		88	
Gross energy (kcal/kg)		4 150	60	4 180	60
Nitrogen-free extract		30.3		28.3	
Starch–acid hydrolysis		3.3		3.0	
Starch–enzymic hydrolysis					
Sugars		10.0		8.20	
Ether extract		1.8	0.5	2.0	0.5
Crude fibre		7.4	1.5	5.6	1.3
Acid detergent fibre (ADF)		9.6		8.2	
Neutral detergent figre (NDF)		13.5		12.3	
Crude protein		**42.5**	**1.5**	**45.8**	**0.8**
Amino acids:					
Lysine		2.70		2.91	0.09
Methionine		0.59		0.63	
Methionine + cystine		1.27		1.37	
Tryptophan		0.57		0.62	
Threonine		1.67		1.79	
Glycine + serine		3.94		4.25	
Leucine		3.26		3.50	
Isoleucine		2.14		2.30	
Valine		2.18		2.35	
Histidine		1.05		1.14	
Arginine		3.18		3.43	
Phenylalanine + tyrosine		3.65		3.93	
Ash		6.00	0.50	6.30	
Calcium		0.30		0.30	
Total phosphorus		0.62		0.69	
Sodium		0.01		0.01	
Potassium		1.70		2.10	
Chloride		traces		traces	
Magnesium		0.25		0.28	
Pig					
Digestible energy — DE (kcal/kg)		3 430		3 500	
Metabolizable energy — ME (kcal/kg)		3 135		3 185	
Apparent crude protein digestibility — ACPD (%)		87		88	
Rabbit					
Digestible energy — DE (kcal/kg)		3 260		3 310	
Indigestible crude fibre — IDCF (%)		6.8		4.8	
Apparent crude protein digestibility — ACPD (%)		84		84	
Poultry					
Metabolizable energy — ME (kcal/kg):					
Broiler					
Pullet and laying hen		2 420		2 500	100
Adult bird		2 250		2 325	100
Apparent crude protein digestibility — ACPD (%)		83			
Available phosphorus (%)		**0.10**		**0.10**	

Table 93 (cont.)

Oil meals	Code:	133		134
Characteristics (% fresh weight)		Soya-bean meal 50		Sunflower seed meal
		Mean	SE	
Dry matter		88		90
Gross energy (kcal/kg)		4 200		4 100
Nitrogen-free extract		28.5		26.0
Starch–acid hydrolysis		3.1	0.5	5.0
Starch–enzymic hydrolysis				
Sugars		8.7	0.2	2.0
Ether extract		1.9	0.5	1.8
Crude fibre		3.4	0.2	26.5
Acid detergent fibre (ADF)		5.3		32.0
Neutral detergent figre (NDF)		7.5		45.0
Crude protein		**48.0**	**1.1**	**29.5**
Amino acids:				
Lysine		3.05	0.11	1.07
Methionine		0.66		0.73
Methionine + cystine		1.43		1.26
Tryptophan		0.65		0.38
Threonine		1.88		1.06
Glycine + serine		4.45		2.91
Leucine		3.68		1.89
Isoleucine		2.42		1.40
Valine		2.46		1.63
Histidine		1.19		0.72
Arginine		3.59		2.56
Phenylalanine + tyrosine		4.12		2.18
Ash		**6.20**	**0.45**	**6.22**
Calcium		0.27		0.35
Total phosphorus		0.69		0.90
Sodium		0.01		0.03
Potassium		2.20		1.10
Chloride		traces		0.11
Magnesium		0.28		0.50
Pig				
Digestible energy — DE (kcal/kg)		3 570		2 200
Metabolizable energy — ME (kcal/kg)		3 235		2 000
Apparent crude protein digestibility — ACPD (%)		89		77
Rabbit				
Digestible energy — DE (kcal/kg)		3 410		2 770
Indigestible crude fibre — IDCF (%)		3.0		18.6
Apparent crude protein digestibility — ACPD (%)		85		86
Poultry				
Metabolizable energy — ME (kcal/kg):				
Broiler		2 540	110	1 385
Pullet and laying hen		2 560	110	1 515
Adult bird		2 375	110	1 415
Apparent crude protein digestibility — ACPD (%)		83		83
Available phosphorus (%)		**0.10**		**0.15**

Table 93 (cont.)

Microbial proteins/algae	Code:	135	136	137	138
Characteristics (% fresh weight)			Algae		Brewer's yeast
		Spirulina	Chlorella	Scenedesmus	
Dry matter		93	93	93	93
Gross energy (kcal/kg)		4 740		3 700	4 300
Nitrogen-free extract		12.3	31.0	15.7	32.7
Starch–acid hydrolysis					4.9
Starch–enzymic hydrolysis					
Sugars					6.9
Ether extract		5.5	6.0	6.5	1.8
Crude fibre		4.4	6.0	11.0	2.8
Acid detergent fibre (ADF)					
Neutral detergent figre (NDF)					
Crude protein		**61.7**	**45.0**	**51.0**	**48.4**
Amino acids:					
Lysine		2.60	3.10	2.70	3.38
Methionine		1.20	0.70	0.70	0.67
Methionine + cystine		1.70	1.00	1.30	1.19
Tryptophan		0.90	0.90	0.80	0.55
Threonine		2.80	1.80	2.40	2.21
Glycine + serine		5.50	3.00	3.50	4.12
Leucine		5.00	3.20	3.80	3.11
Isoleucine		3.50	1.20	1.70	2.15
Valine		3.90	2.10	2.90	2.48
Histidine		1.00	1.30	1.00	1.12
Arginine		4.30	2.00	0.50	2.27
Phenylalanine + tyrosine		5.20	4.40	3.50	3.36
Ash		8.30	5.0	6.80	7.25
Calcium		0.1	0.1	1.5	0.14
Total phosphorus		1.0	1.4	0.4	1.40
Sodium		1.4		0.4	0.07
Potassium		1.5		1.7	1.70
Chloride		1.2			0.15
Magnesium		0.15		1.50	0.20
Pig					
Digestible energy — DE (kcal/kg)					3 530
Metabolizable energy — ME (kcal/kg)					3 155
Apparent crude protein digestibility — ACPD (%)			76	50	88
Rabbit					
Digestible energy — DE (kcal/kg)					2 990
Indigestible crude fibre — IDCF (%)					2.20
Apparent crude protein digestibility — ACPD (%)					85
Poultry					
Metabolizable energy — ME (kcal/kg):					
Broiler					
Pullet and laying hen					2 480
Adult bird					2 290
Apparent crude protein digestibility — ACPD (%)					
Available phosphorus (%)		**0.40**			**0.91**

Table 93 (cont.)

Microbial proteins/algae	Code:	139	140	141	142	143
Characteristics (% fresh weight)			Yeast		Fodder protein	ICI Pruteen
		Distiller's	Torula	Lactic		
Dry matter		92	92	95	91	90
Gross energy (kcal/kg)		4 250	4 295		4 660	4 790
Nitrogen-free extract		37.1	33.1	30.3	11.1	
Starch–acid hydrolysis						1.0
Starch–enzymic hydrolysis				6.8	7.4	
Sugars					1.1	
Ether extract		1.9	2.0	6.2	5.0	7.8
Crude fibre		3.0	1.8		0.3	0.9
Acid detergent fibre (ADF)						
Neutral detergent figre (NDF)						
Crude protein		**41.8**	**47.8**	**50.7**	**70.0**	**74.0**
Amino acids:						
Lysine		3.15	3.47	3.70	2.45	4.60
Methionine		0.65	0.70	0.66	0.98	1.80
Methionine + cystine		1.11	1.14	1.16	1.18	2.30
Tryptophan		0.46	0.50	0.55	0.40	1.00
Threonine		2.04	2.31	2.99	2.47	3.50
Glycine + serine		3.55	4.15	4.80	4.61	6.30
Leucine		2.85	3.33	3.80	3.96	5.30
Isoleucine		2.09	2.63	2.28	2.55	3.50
Valine		2.47	2.75	2.68	3.36	4.20
Histidine		0.90	0.98	0.96	1.01	1.30
Arginine		2.00	2.63	2.43	3.26	3.50
Phenylalanine + tyrosine		3.11	3.73	3.74	3.50	5.00
Ash		**8.20**	**7.31**	**7.80**	**4.55**	**8.20**
Calcium		0.35	0.55	0.23	0.16	1.30
Total phosphorus		1.50	1.50	1.80	0.70	2.10
Sodium		0.07	0.01	0.12		0.80
Potassium		1.80	1.60	2.20		0.18
Chloride		0.18	0.20	0.22	0.44	0.03
Magnesium		0.20	0.12	0.14		0.22
Pig						
Digestible energy — DE (kcal/kg)		3 440	3 435	3 590	3 870	
Metabolizable energy — ME (kcal/kg)		3 170	3 115	3 250	3 515	
Apparent crude protein digestibility — ACPD (%)		80	85	85	65	88
Rabbit						
Digestible energy — DE (kcal/kg)						
Indigestible crude fibre — IDCF (%)						
Apparent crude protein digestibility — ACPD (%)						
Poultry						
Metabolizable energy — ME (kcal/kg):						
Broiler						3 690
Pullet and laying hen						3 590
Adult bird						3 300
Apparent crude protein digestibility — ACPD (%)						
Available phosphorus (%)		**1.00**	**1.00**	**1.20**	**0.63**	**1.40**

Table 93 (cont.)

Animal by-products	Code:	144	145	146
Characteristics (% fresh weight)		Fish protein concentrate	Krill	
			Concentrate	Meal
Dry matter		95	93	88
Gross energy (kcal/kg)		5 140		
Nitrogen-free extract			1.6	7.2
Starch–acid hydrolysis				
Starch–enzymic hydrolysis				
Sugars				
Ether extract		10.5	0.3	9.2
Crude fibre			6.0*	5.7**
Acid detergent fibre (ADF)				
Neutral detergent figre (NDF)				
Crude protein		**83.3**	**72.1**	**53.0**
Amino acids:				
Lysine		5.67	3.00	2.17
Methionine		2.48	2.02	1.50
Methionine + cystine		3.25	3.10	2.10
Tryptophan		0.62	0.94	0.68
Threonine		3.16	3.61	2.38
Glycine + serine			7.28	4.95
Leucine		4.98	5.98	4.45
Isoleucine		3.00	4.28	3.15
Valine		3.56	4.01	2.95
Histidine		1.68	1.80	1.28
Arginine		5.38	4.43	3.26
Phenylalanine + tyrosine		5.24	7.93	4.80
Ash		**7.10**	**13.00**	**12.90**
Calcium		0.60	3.20	3.20
Total phosphorus		0.80	2.00	2.00
Sodium			1.70	1.70
Potassium			1.30	1.30
Chloride				
Magnesium			0.60	0.60
Pig				
Digestible energy — DE (kcal/kg)			3 740	3 580
Metabolizable energy — ME (kcal/kg)			3 230	3 225
Apparent crude protein digestibility — ACPD (%)			94	86
Rabbit				
Digestible energy — DE (kcal/kg)				
Indigestible crude fibre — IDCF (%)				
Apparent crude protein digestibility — ACPD (%)				
Poultry				
Metabolizable energy — ME (kcal/kg):				
Broiler			3 430	3 380
Pullet and laying hen			3 360	3 330
Adult bird			3 160	3 180
Apparent crude protein digestibility — ACPD (%)				
Available phosphorus (%)		**0.68**	**1.70**	**1.70**

*Chitin 5.3%
**Chitin 2.5%

Table 93 (cont.)

Animal by-products	Code:	147	148	149	150	151
Characteristics (% fresh weight)		\multicolumn{5}{c}{Fish meal}					
		\multicolumn{3}{c}{Crude protein}	\multicolumn{2}{c}{Defatted crude protein}
		60%	65%	72%	65%	72%
Dry matter		92	92	92	92	92
Gross energy (kcal/kg)		4 530	4 780	4 940	4 600	4 300
Nitrogen-free extract						
Starch–acid hydrolysis						
Starch–enzymic hydrolysis						
Sugars						
Ether extract		9.2	9.6	9.5	5.5	1.8
Crude fibre						
Acid detergent fibre (ADF)						
Neutral detergent figre (NDF)						
Crude protein		**59.3**	**66.2**	**71.6**	**64.6**	**71.3**
Amino acids:						
Lysine		4.41	5.03	5.48	5.04	5.42
Methionine		1.62	1.92	2.08	1.81	2.07
Methionine + cystine		2.19	2.52	2.80	2.39	2.71
Tryptophan		0.63	0.70	0.79	0.65	0.75
Threonine		2.55	2.88	3.06	2.73	3.10
Glycine + serine		6.70	6.62	7.12	5.59	7.13
Leucine		4.44	5.10	5.20	4.81	5.20
Isoleucine		2.80	3.11	3.42	3.04	3.35
Valine		3.19	3.67	3.88	3.55	3.80
Histidine		1.37	1.52	1.61	1.58	1.63
Arginine		3.46	3.77	3.92	3.71	4.06
Phenylalanine + tyrosine		4.15	4.83	4.85	4.81	5.00
Ash		**20.70**	**15.60**	**11.00**	**21.40**	**16.80**
Calcium		6.20	3.90	2.70	6.30	4.20
Total phosphorus		3.40	2.55	1.80	3.50	2.75
Sodium		1.00	0.90	0.77	1.00	0.95
Potassium		0.70	0.75	0.90	0.70	0.74
Chloride		1.30	1.15	1.00	1.30	1.20
Magnesium		0.23	0.20	0.18	0.23	0.21
Pig						
Digestible energy — DE (kcal/kg)		3 650	3 850	4 400	3 700	3 800
Metabolizable energy — ME (kcal/kg)		3 240	3 415	3 885	3 255	3 295
Apparent crude protein digestibility — ACPD (%)		90	86	94	90	94
Rabbit						
Digestible energy — DE (kcal/kg)						
Indigestible crude fibre — IDCF (%)						
Apparent crude protein digestibility — ACPD (%)						
Poultry						
Metabolizable energy — ME (kcal/kg):						
Broiler		3 200	3 535	3 790	3 180	3 280
Pullet and laying hen		3 135	3 460	3 700	3 110	3 135
Adult bird		2 935	3 235	3 440	2 905	2 930
Apparent crude protein digestibility — ACPD (%)						
Available phosphorus (%)		**2.90**	**2.20**	**1.53**	**3.00**	**2.35**

Table 93 (cont.)

Animal by-products	Code:	152	153	154	155	156
Characteristics (% fresh weight)		Fish solubles	Greaves	Feather meal	Blood meal	Meatmeal 50% crude protein
Dry matter		93	93	93	90	93
Gross energy (kcal/kg)		4 515	4 730	5 115	4 950	3 815
Nitrogen-free extract						
Starch–acid hydrolysis						
Starch–enzymic hydrolysis						
Sugars						
Ether extract		9.3	10.1	3.5	1.1	10.0
Crude fibre						
Acid detergent fibre (ADF)						
Neutral detergent figre (NDF)						
Crude protein		**63.3**	**67.2**	**85.8**	**84.0**	**50.5**
Amino acids:						
Lysine		2.91	3.72	1.84	7.62	2.83
Methionine		0.93	0.95	0.53	0.93	0.71
Methionine + cystine		1.36	1.57	4.08	1.68	1.21
Tryptophan		0.47	0.38	0.43	1.06	0.29
Threonine		1.33	2.27	3.91	4.00	1.67
Glycine + serine		7.40	12.94	17.57	8.80	9.26
Leucine		2.78	4.00	7.14	11.58	3.13
Isoleucine		1.55	1.96	3.95	0.79	1.41
Valine		1.90	3.10	6.88	7.02	2.33
Histidine		2.22	1.12	0.58	5.18	0.96
Arginine		2.69	4.42	5.66	3.63	3.52
Phenylalanine + tyrosine		2.22	3.74	7.01	6.97	2.78
Ash		**17.80**	**15.00**	**3.17**	**4.45**	**30.30**
Calcium		0.17	4.10	0.20	0.30	9.30
Total phosphorus		1.00	2.10	0.70	0.25	4.50
Sodium		1.5–3.0	1.00		0.32	0.70
Potassium		2.50	0.40	0.24	0.10	0.50
Chloride		3.5–7.0			0.40	0.60
Magnesium		0.02	0.12	0.18	0.22	0.70
Pig						
Digestible energy — DE (kcal/kg)		3 700	4 280	3 730	3 900	2 550
Metabolizable energy — ME (kcal/kg)		3 270	3 825	3 240	3 415	2 250
Apparent crude protein digestibility — ACPD (%)		89	88	75	85	78
Rabbit						
Digestible energy — DE (kcal/kg)						
Indigestible crude fibre — IDCF (%)						
Apparent crude protein digestibility — ACPD (%)						
Poultry						
Metabolizable energy — ME (kcal/kg):						
Broiler			3 470	2 880	3 190	2 790
Pullet and laying hen			3 420	2 800	3 150	2 755
Adult bird		3 030	3 260	2 565	2 820	2 560
Apparent crude protein digestibility — ACPD (%)						
Available phosphorus (%)		0.93	1.80	0.60	0.22	3.60

Table 93 (*cont.*)

Animal by-products	Code:	157	158	159	160
Characteristics (% fresh weight)			Meat meal		
		55% Crude protein	Defatted crude protein		
			50%	55%	60%
Dry matter		93	93	93	93
Gross energy (kcal/kg)		4 070	3 450	3 400	4 040
Nitrogen-free extract					
Starch–acid hydrolysis					
Starch–enzymic hydrolysis					
Sugars					
Ether extract		9.9	5.0	5.0	4.0
Crude fibre					
Acid detergent fibre (ADF)					
Neutral detergent figre (NDF)					
Crude protein		53.8	50.0	56.1	60.0
Amino acids:					
Lysine		2.82	2.80	2.95	3.46
Methionine		0.74	0.70	0.77	0.85
Methionine + cystine		1.24	1.20	1.29	1.39
Tryptophan		0.28	0.28	0.29	0.36
Threonine		1.73		1.80	2.15
Glycine + serine		9.20	9.17	9.60	10.10
Leucine		3.17	3.10	3.31	3.82
Isoleucine		1.57	1.40	1.64	1.94
Valine		2.31	2.30	2.41	2.80
Histidine		0.88	0.95	0.92	1.13
Arginine		3.49	3.48	3.64	3.89
Phenylalanine + tyrosine		2.89	2.75	3.01	3.36
Ash		25.00	32.60	28.80	22.50
Calcium		7.70	10.20	9.00	7.05
Total phosphorus		3.70	4.85	4.30	3.35
Sodium		0.58	0.76	0.67	0.52
Potassium		0.41	0.54	0.48	0.37
Chloride		0.50	0.65	0.57	0.45
Magnesium		0.58	0.76	0.67	0.52
Pig					
Digestible energy — DE (kcal/kg)		3 000	2 600	2 670	3 150
Metabolizable energy — ME (kcal/kg)		2 670	2 300	2 320	2 775
Apparent crude protein digestibility — ACPD (%)		80	79	82	82
Rabbit					
Digestible energy — DE (kcal/kg)					
Indigestible crude fibre — IDCF (%)					
Apparent crude protein digestibility — ACPD (%)					
Poultry					
Metabolizable energy — ME (kcal/kg):					
Broiler		3 035	2 310	2 535	2 750
Pullet and laying hen		2 980	2 260	2 475	2 685
Adult bird		2 810	2 100	2 300	2 500
Apparent crude protein digestibility — ACPD (%)					
Available phosphorus (%)		3.00	3.90	3.43	2.70

Table 93 (cont.)

Animal by-products Characteristics (% fresh weight)	Code:	161 Meat and bone meal	162 Poultry litter	163 Poultry offal meal	164 Poultry hatchery waste
Dry matter		93	88	93	94
Gross energy (kcal/kg)		3 160		4 700	
Nitrogen-free extract					
Starch–acid hydrolysis					
Starch–enzymic hydrolysis					
Sugars					
Ether extract		7.2	3.1	12.5	12.2
Crude fibre			13.7		
Acid detergent fibre (ADF)					
Neutral detergent figre (NDF)					
Crude protein		**42.7**	**25.0**	**58.0**	**27.2**
Amino acids:					
Lysine		2.11	0.35	2.33	1.37
Methionine		0.53	0.21	0.84	0.64
Methionine + cystine		0.91	0.40	2.32	1.20
Tryptophan		0.21		0.39	0.35
Threonine		1.32	0.36	2.13	1.03
Glycine + serine		8.00		9.20	3.25
Leucine		2.14		3.80	2.25
Isoleucine		1.08		2.40	1.31
Valine		1.77		3.06	1.61
Histidine		0.64		0.70	0.50
Arginine		2.90		3.38	1.76
Phenylalanine + tyrosine		2.18		3.82	1.71
Ash		41.10	23.60	18.50	42.20*
Calcium		12.90	6.00	4.00	16.50
Total phosphorus		6.10	2.15	2.35	0.52
Sodium		0.95	0.70	0.70	0.33
Potassium		0.68	2.30	0.70	0.33
Chloride		0.80	1.00	0.70	0.30
Magnesium		0.95	1.00	0.15	0.27
Pig					
Digestible energy — DE (kcal/kg)				4 090	
Metabolizable energy — ME (kcal/kg)				3 705	
Apparent crude protein digestibility — ACPD (%)				85	
Rabbit					
Digestible energy — DE (kcal/kg)					
Indigestible crude fibre — IDCF (%)					
Apparent crude protein digestibility — ACPD (%)					
Poultry					
Metabolizable energy — ME (kcal/kg):					
Broiler		1 975			
Pullet and laying hen		1 930		3 030	1 775
Adult bird		1 820		2 800	1 670
Apparent crude protein digestibility — ACPD (%)					
Available phosphorus (%)		**4.90**		**2.00**	**0.45**

*Corresponding to approximately 43% from shells and 57% from egg contents.

Table 93 (cont.)

Dairy products — liquid	Code:	165	166	167	168	169
		Milk		Sweet butter-milk	Separated whey	
Characteristics (% fresh weight)		Whole	Skimmed		Sweet	Acid
Dry matter		12.5	9.1	9.5	6.4	6.2
Gross energy (kcal/kg)		750	395	410	240	240
Nitrogen-free extract		4.8	4.8	4.9	5.0	4.6
Starch–acid hydrolysis						
Starch–enzymic hydrolysis						
Sugars		4.8	4.8	4.8	4.7	4.2
Ether extract		3.50	0.09	0.25	0.05	0.05
Crude fibre						
Acid detergent fibre (ADF)						
Neutral detergent figre (NDF)						
Crude protein		**3.4**	**3.5**	**3.5**	**0.82**	**0.80**
Amino acids:						
Lysine		0.26	0.27	0.29	0.07	0.07
Methionine		0.08	0.07	0.09	0.01	0.02
Methionine + cystine		0.14	0.10	0.12	0.03	0.04
Tryptophan		0.05	0.05	0.05	0.01	0.01
Threonine		0.16	0.15	0.15	0.05	0.05
Glycine + serine		0.27	0.27	0.27	0.06	0.05
Leucine		0.34	0.34	0.35	0.08	0.07
Isoleucine		0.22	0.19	0.19	0.05	0.04
Valine		0.23	0.23	0.25	0.05	0.04
Histidine		0.09	0.09	0.10	0.02	0.02
Arginine		0.12	0.12	0.11	0.02	0.02
Phenylalanine + tyrosine		0.34	0.33	0.36	0.05	0.04
Ash		**0.80**	**0.70**	**0.80**	**0.52**	**0.71**
Calcium		0.13	0.12	0.12	0.04	0.12
Total phosphorus		0.10	0.10	0.08	0.04	0.06
Sodium		0.05	0.06	0.09	0.04	0.05
Potassium		0.15	0.15	0.16	0.15	0.15
Chloride		0.10	0.10	0.10	0.08	0.10
Magnesium		0.01	0.01	0.01	0.01	0.01
Pig						
Digestible energy — DE (kcal/kg)		700	380	395	225	210
Metabolizable energy — ME (kcal/kg)		670	350	365	220	205
Apparent crude protein digestibility — ACPD (%)		95	95	95	95	95
Rabbit						
Digestible energy — DE (kcal/kg)						
Indigestible crude fibre — IDCF (%)						
Apparent crude protein digestibility — ACPD (%)						
Poultry						
Metabolizable energy — ME (kcal/kg):						
Broiler						
Pullet and laying hen						
Adult bird						
Apparent crude protein digestibility — ACPD (%)						
Available phosphorus (%)						

Table 93 (*cont.*)

Dairy products — liquid	Code:	170	171	172	173
Characteristics *(% fresh weight)*		Mixed separated whey	Acid goat whey	Whey casein	
				Rennet	Lactic acid
Dry matter		6.5	6.1	6.9	6.5
Gross energy (kcal/kg)		240			
Nitrogen-free extract		5.0	5.3	5.2	5.0
Starch–acid hydrolysis					
Starch–enzymic hydrolysis					
Sugars					
Ether extract		4.5	3.8	4.8	4.5
Crude fibre					
Acid detergent fibre (ADF)					
Neutral detergent figre (NDF)					
Crude protein		**0.9**	**0.9**	**0.9**	**0.6**
Amino acids:					
Lysine		0.07	0.06	0.07	0.07
Methionine		0.01	0.01	0.01	0.02
Methionine + cystine		0.03	0.03	0.03	0.04
Tryptophan		0.01	0.01	0.01	0.01
Threonine		0.05	0.05	0.05	0.05
Glycine + serine		0.06	0.06	0.06	0.05
Leucine		0.07	0.07	0.08	0.07
Isoleucine		0.05	0.05	0.05	0.04
Valine		0.04	0.05	0.05	0.04
Histidine		0.02	0.02	0.02	0.02
Arginine		0.02	0.02	0.02	0.02
Phenylalanine + tyrosine		0.05	0.05	0.05	0.04
Ash		**0.60**	**0.81**	**0.45**	**0.75**
Calcium		0.06	0.13	0.03	0.12
Total phosphorus		0.05	0.07	0.03	0.06
Sodium		0.06	0.04	0.03	0.05
Potassium		0.15	0.17	0.13	0.15
Chloride		0.08	0.11	0.07	0.10*
Magnesium		0.01		0.01	0.01
Pig					
Digestible energy — DE (kcal/kg)		220		220	220
Metabolizable energy — ME (kcal/kg)		210		210	210
Apparent crude protein digestibility — ACPD (%)		95		95	95
Rabbit					
Digestible energy — DE (kcal/kg)					
Indigestible crude fibre — IDCF (%)					
Apparent crude protein digestibility — ACPD (%)					
Poultry					
Metabolizable energy — ME (kcal/kg):					
Broiler					
Pullet and laying hen					
Adult bird					
Apparent crude protein digestibility — ACPD (%)					
Available phosphorus (%)					

*1.60% in casein hydrochloride.

Table 93 (cont.)

Dairy products — liquid	Code:	174	175	176	177	178
Characteristics (% fresh weight)		Milk		Sweet butter-milk	Casein	
		Whole	Skimmed		Lactic acid	Hydro-chloric acid
Dry matter		97	95	95	91	91
Gross energy (kcal/kg)		5 640	4 120	4 100	5 220	5 200
Nitrogen-free extract		39.5	51.7	47.2	4.8	6.1
Starch–acid hydrolysis						
Starch–enzymic hydrolysis						
Sugars		39.5	51.0	41.5		
Ether extract		26.0	0.8	5.2	1.5	1.4
Crude fibre						
Acid detergent fibre (ADF)						
Neutral detergent figre (NDF)						
Crude protein		**26.0**	**34.9**	**34.5**	**82.0**	**83.0**
Amino acids:						
Lysine		2.18	2.81	2.84	6.72	6.89
Methionine		0.68	0.85	0.86	2.29	2.86
Methionine + cystine		0.91	1.25	1.17	2.57	3.27
Tryptophan		0.39	0.44	0.48	1.39	
Threonine		1.19	1.53	1.45	4.02	3.93
Glycine + serine		2.00	2.38	2.66	7.37	6.97
Leucine		2.60	3.42	3.46	7.54	8.33
Isoleucine		1.48	2.12	1.87	5.00	4.85
Valine		1.84	2.30	2.42	5.91	6.54
Histidine		0.75	0.99	0.97	2.54	2.48
Arginine		0.91	1.20	1.11	3.36	3.31
Phenylalanine + tyrosine		2.70	3.24	3.60	9.26	9.77
Ash		**6.00**	**7.60**	**8.11**	**2.70**	**0.50**
Calcium		0.91	1.30	1.22	0.12	0.02
Total phosphorus		0.63	1.00	0.85	0.41	0.04
Sodium		0.47	0.50	0.90	0.004	0.002
Potassium		1.72	1.65	1.60	0.006	0.003
Chloride		0.71	0.75	1.00	0.04	0.21
Magnesium		0.08	0.11	0.10	0.002	
Pig						
Digestible energy — DE (kcal/kg)		5 240	3 900	3 950	4 210	4 200
Metabolizable energy — ME (kcal/kg)		5 015	3 630	3 675	3 570	3 550
Apparent crude protein digestibility — ACPD (%)		95	95	95	95	95
Rabbit						
Digestible energy — DE (kcal/kg)						
Indigestible crude fibre—IDCF (%)						
Apparent crude protein digestibility — ACPD (%)						
Poultry						
Metabolizable energy — ME (kcal/kg):						
Broiler			3 300		3 900	
Pullet and laying hen			3 260		3 800	
Adult bird			3 130		3 500	
Apparent crude protein digestibility — ACPD (%)						
Available phosphorus (%)						

Table 93 (*cont.*)

Dairy products — dried *Characteristics* *(% fresh weight)*	Code:	179 Sodium caseinate	180 Separated whey Sweet	181 Mixed	182 Sweet ewe whey
Dry matter		97	96	95	96
Gross energy (kcal/kg)		5 380	3 600	3 620	3 850
Nitrogen-free extract		2.9	74.5	71.6	65.9
Starch–acid hydrolysis					
Starch–enzymic hydrolysis					
Sugars		2.1	71.0	66.5	62.0
Ether extract		2.0	0.9	1.3	1.2
Crude fibre					
Acid detergent fibre (ADF)					
Neutral detergent figre (NDF)					
Crude protein		**88.6**	**12.8**	**13.3**	**22.0**
Amino acids:					
Lysine		7.39	1.13	1.12	1.80
Methionine		2.67	0.22	0.22	0.35
Methionine + cystine		3.07	0.53	0.50	0.85
Tryptophan			0.20	0.15	0.40
Threonine		3.78	0.53	0.72	1.35
Glycine + serine		6.73	0.91	0.81	1.40
Leucine		8.47	1.31	1.14	1.80
Isoleucine		4.80	0.74	0.72	1.25
Valine		5.65	0.76	0.67	1.30
Histidine		2.81	0.25	0.24	0.50
Arginine		3.52	0.37	0.33	0.40
Phenylalanine + tyrosine		9.83	0.84	0.77	1.15
Ash		3.52	7.80	8.80	6.90
Calcium		0.13	0.62	0.94	0.61
Total phosphorus		0.77	0.66	0.70	0.67
Sodium		1.32	0.58	0.80	0.75
Potassium		0.35	2.20	2.23	1.57
Chloride			1.54	1.20	1.00
Magnesium			0.12	0.12	
Pig					
Digestible energy — DE (kcal/kg)		4 840	3 350	3 210	3 560
Metabolizable energy — ME (kcal/kg)			3 230	3 090	3 420
Apparent crude protein digestibility — ACPD (%)		95	95	95	
Rabbit					
Digestible energy — DE (kcal/kg)					
Indigestible crude fibre—IDCF (%)					
Apparent crude protein digestibility — ACPD (%)					
Poultry					
Metabolizable energy — ME (kcal/kg):					
Broiler					
Pullet and laying hen					
Adult bird					
Apparent crude protein digestibility — ACPD (%)					
Available phosphorus (%)					

Table 93 (*cont.*)

Dairy products — dried *Characteristics* *(% fresh weight)*	Code:	183	184	185	186
		Permeate of:		*Partially delactosed whey**	*Whey proteins*
		Milk	Mixed whey		
Dry matter		99	99	95	93
Gross energy (kcal/kg)		3 500	3 490	3 120	
Nitrogen-free extract		89.2	86.4	54.0	
Starch–acid hydrolysis					
Starch–enzymic hydrolysis					
Sugars		86	82	40	10
Ether extract		0	0	2	
Crude fibre					
Acid detergent fibre (ADF)					
Neutral detergent figre (NDF)					
Crude protein		2.8	3.7	17.0	84.0
Amino acids:					
Lysine				0.83	7.40
Methionine				0.34	1.40
Methionine + cystine					3.50
Tryptophan					1.30
Threonine				1.65	3.50
Glycine + serine				1.60	6.00
Leucine				1.25	6.60
Isoleucine				1.40	4.90
Valine				1.25	5.00
Histidine				0.31	1.60
Arginine				0.31	2.40
Phenylalanine + tyrosine				0.75	5.50
Ash		7.00	8.90	22.00	3.00
Calcium			0.60	2.00	
Total phosphorus			0.59	1.20	
Sodium			0.76	2.50	
Potassium			2.57	6.30	
Chloride			0.14		
Magnesium			1.15	1.00	
Pig					
Digestible energy — DE (kcal/kg)				2 800	
Metabolizable energy — ME (kcal/kg)				2 670	
Apparent crude protein digestibility — ACPD (%)				85	96
Rabbit					
Digestible energy — DE (kcal/kg)					
Indigestible crude fibre—IDCF (%)					
Apparent crude protein digestibility — ACPD (%)					
Poultry					
Metabolizable energy — ME (kcal/kg):					
Broiler					
Pullet and laying hen					
Adult bird					
Apparent crude protein digestibility — ACPD (%)					
Available phosphorus (%)					

*Source of wheys

Table 93 (cont.)

Amino acids	Code: 187	188	189
Characteristics (% fresh weight)	dl-Methionine**	l-Lysine hydro-chloride**	Methionine hydroxyanalogue*
Dry matter	100	98	98
Gross energy (kcal/kg)	5 750	4 970	4 190
Nitrogen-free extract			
Starch–acid hydrolysis			
Starch–enzymic hydrolysis			
Sugars			
Ether extract			
Crude fibre			
Acid detergent fibre (ADF)			
Neutral detergent figre (NDF)			
Crude protein†	58.7	95.6	0
Amino acids:			
Lysine	0	79	
Methionine	99	0	75
Methionine + cystine	99	0	75
Tryptophan			
Threonine			
Glycine + serine			
Leucine			
Isoleucine			
Valine			
Histidine			
Arginine			
Phenylalanine + tyrosine			
Ash	0.2	0	20
Calcium	0.02	0.04	9.80
Total phosphorus			
Sodium			
Potassium	0.04	0.03	0.02
Chloride		19.4	
Magnesium			
Pig			
Digestible energy — DE (kcal/kg)	5 750	4 970	4 190
Metabolizable energy — ME (kcal/kg)	5 280	4 250	3 830
Apparent crude protein digestibility — ACPD (%)			
Rabbit			
Digestible energy — DE (kcal/kg)	5 750	4 970	4 190
Indigestible crude fibre—IDCF (%)			
Apparent crude protein digestibility — ACPD (%)			
Poultry			
Metabolizable energy — ME (kcal/kg):			
Broiler	5 020	3 990	3 610
Pullet and laying hen	4 950	3 870	3 550
Adult bird	4 730	3 510	3 370
Apparent crude protein digestibility — ACPD (%)			
Available phosphorus (%)			

For footnotes *see* opposite.

Table 93 (cont.)

Sources of calcium

Characteristics (% fresh weight)	Code: 190 Calcium carbonate	191 Lime-stone	192 Sea shells	193 Carbonate from sugar beet factories	194 Oyster shells	195 Sea shells
Formula	–	–	–	–	–	–
Crude protein	0	0	0	2.8	–	–
Calcium	38	38–39	33	31.5	38	35
Total phosphorus	0	0.02	0.3	0.4	0.05	0.03
Sodium	0.02	0.06	0.55	–	0.3	0.1
Potassium	–	–	0.11	–	0.15	–
Chloride	–	–	–	–	0.01	0.14
Magnesium	–	0–0.3	0.3	2.5	0.3	1.2
Other important elements (in mg/kg unless otherwise stated)	–	Fe: 70–700 Mn: 150 Se: 0.15	Mn: 150	Fe: 900 Mn: 120 Zn: 40 Cu: 30 Co: 20	Fe: 200–700 Mn: 100–400	–

Sources of calcium

Characteristics (% fresh weight)	Code: 196 Egg shells Dried	197 Egg shells Clean	198 Marl	199 Hatchery ashes	200 Portland cement	201 Gypsum
Formula	–	–	–	–	–	–
Crude protein	7.5	5.3	0.35	–	–	–
Calcium	35	36	32.7	37.6	45.7	23–29
Total phosphorus	0.12	0.11	0.05	1.2	–	–
Sodium	0.15	0.12	0.5	–	0.2	–
Potassium	0.1	0.06	0.04	–	–	–
Chloride	–	–	0.5	–	–	–
Magnesium	0.4	0.4	2.3–4.4	–	1.2	–
Other important elements (in mg/kg unless otherwise stated)	Fe: 20	Fe: 20	Fe: 0.5–1.6% Mn: 200–800 Al: 900–9 000 Sr: 2 000	–	Fe: 100 Mn: 2 000 Al: 300 Si: 10%	SO_4: 66%

*Although containing no nitrogen, methionine hydroxyanalogue promotes nitrogen retention by combining with amine groups that would otherwise be lost from the animal.
**A complete availability of synthetic amino acids, when compared to those found in other raw materials, may be assumed and a figure higher than their chemical value may be given. In this case, an availability of 110% and 92% for dl-methionine and lysine hydrochloride, respectively, may be used.
†Values obtained by multiplying the figure for crude protein by 6.25.

Table 93 (*cont.*)

Sources of water-soluble phosphorus

Characteristics (%fresh weight)	Code: 202 Phosphoric acid	203 Monosodium phosphate Di-hydrate	204 Monosodium phosphate Anhydrous	205 Disodium phosphate Di-hydrate	206 Disodium phosphate Anhydrous	207 Mono-potassium phosphate
Formula	H_3PO_4	$NaH_2PO_4 \cdot 2H_2O$	NaH_2PO_4	$Na_2HPO_4 \cdot 2H_2O$	Na_2HPO_4	KH_2PO_4
Crude protein	–	–	–	–	–	–
Calcium	–	–	–	8.5	–	–
Total phosphorus	24–28	19.8	25	12.5	21.3	22.6
Sodium	0.03	14.5	19	–	32	–
Potassium	–	–	–	–	–	28
Chloride	–	–	0.15	–	–	–
Magnesium	–	–	–	–	–	–
Other important elements (in mg/kg unless otherwise stated)						

Sources of water-soluble phosphorus

Characteristics (% fresh weight)	Code: 208 Dipotassium phosphate	209 Mono-ammonium phosphate	210 Di-ammonium phosphate	211 Sodium tripoly-phosphate	212 Monocalcium phosphate
Formula	K_2HPO_4	$NH_4H_2PO_4$	$(NH_4)_2HPO_4$	$Na_5P_3O_{10}$	$Ca(H_2PO_4)_2 \cdot H_2O$
Crude protein	–	–	–	–	–
Calcium	–	–	–	–	17–20
Total phosphorus	17.7	26.7	23.2	24.7	21–24
Sodium	–	–	–	31	0.1
Potassium	44	–	–	–	–
Chloride	–	–	–	–	–
Magnesium	–	–	–	0.06	0.05
Other important elements (in mg/kg unless otherwise stated)		NH_4=15.5%	NH_4=27.3%	–	–

Table 93 (*cont.*)

Sources of water-soluble phosphorus

Characteristics (%fresh weight)	Code:	213 Dicalcium phosphate Hydrated	214 Dicalcium phosphate Anhydrous	215 Mono-dicalcium phosphate	216 Tri-calcium phosphate	217 Rock phosphates Natural	218 Rock phosphates De-fluorinated
Formula		$CaHPO_4 \cdot 2H_2O$	$CaHPO_4$	–	$Ca_3(PO_4)_2$	–	–
Crude protein		–	–	–	–	–	–
Calcium		23–26	28	18–21	37	28–29	32
Total phosphorus		17.5–19	21	20	19.5	13–18	18
Sodium		0.04	0.03	–	–	0.05	4.5
Potassium		–	–	–	–	0.6	–
Chloride		–	–	–	–	–	–
Magnesium		0.01	–	0.3	–	0.2–0.4	0.2
Other important elements (in mg/kg unless otherwise stated)						F: 2%	F: < 2 000 Fe: 7 000 Mn: 700 Se: 1.4 Cu: 70

Sources of water-soluble phosphorus

Characteristics (% fresh weight)	Code:	219 Bonemeal Untreated	220 Bonemeal Degelatinized	221 Ca–Mg–Na phosphate	222 Sodium chloride	223 Sea salt	224 Sodium bicarbonate
Formula		–	–	–	NaCl	NaCl	$NaHCO_3$
Crude protein		6	–	–	–	–	–
Calcium		23.5	30.7	9.5	–	0.8	–
Total phosphorus		11.2	14.1	17.3	–	–	–
Sodium		0.6	0.45	11.5	39.3	35.4	27.0
Potassium		0.23	–	–	–	–	–
Chloride		0.08	0.08	–	60.6	54.5	4
Magnesium		0.24	0.80	5	–	–	–
Other important elements (in mg/kg unless otherwise stated)		ME: 400 kcal/kg Fe: 180 Mn: 3 S: 0.17%	Zn: 420 Cu: 15			Mn: 0.2	HCO_3: 69%

Table 94 Trace element composition of raw materials (mg/kg, fresh weight basis)*

Code		Raw material	Sulphur S	Iron Fe	Copper Cu	Zinc Zn	Manganese Mn	Cobalt Co
	I	**Major cereals**						
1		Oats	2 100	80	5	25	40	0.04
2		Wheat — soft	2 000	55	8	15–50	40	0.02
3		Maize (corn, USA)	1 200	30	3	20	3–10	0.02
4		Barley — 2 row	1 400	80	7	20	15	0.01
5		Barley — 6 row	1 400	80	7	20	15	0.01
6		Sorghum — low tannin	1 000	50	3–14	20	15	0.02
7		Sorghum — high tannin	1 100	50	3–14	20	15	0.02
	II	**Minor cereals**						
8		Oats — naked	–	–	–	30	30	–
9		Oats — dehulled	1 200	40	1	25	35	–
10		Oats — flaked	2 500	55	4	20	40	0.01
11		Wheat — hard	1 700	70	7	15	50	–
12		Millet	1 400	40–120	30	15	30	0.02
13		Barley — naked	1 100	55	7	20	15	–
14		Rice	500	20–80	3	2	20	–
15		Buckwheat	1 800	45	9	9	30	0.05
16		Rye — winter	1 300	70	6	25	60	0.03
17		Triticale — French	–	–	–	30	40	–
	III	**Cereal by-products**						
		From hard wheat						
18		Middlings	2 000	70	7	30	50	–
19		Shorts	–	–	–	–	–	–
20		Bran	2 000	200	15	100	100	–
		From soft wheat						
21		Middlings	2 000	55	5	40	50	–
22		Germ	3 100	80	9	125	140	0.04
23		White shorts	2 000	100	7–20	90	100	0.1
24		Red shorts	1 800	100	5	70	60	0.1
25		Fine bran	1 800	170	10	75	110	0.06
26		Coarse bran	2 100	170	10	75	110	0.06
		From maize (corn, USA)						
27		Distiller's dried grains and solubles (DDGS)	3 700	200	45	70	25	0.1
28		Germ						
29		Gluten feed	1 000	400	10	70	20	0.15
30		Gluten meal 40	6 000	400	25	40	8	–
31		Gluten meal 60	7 200	150	20	40	4	0.1
32		Cobs	4 000	180	6	7	6	–
33		Bran	800	–	–	–	15	–
34		Germ — oilmeal (expeller)	2 000	300	7	70	9	0.1
35		Germ — oilmeal (solvent)	2 800	300	7	80	10	0.1
		From barley						
36		Brewer's grains	3 500	280	20	80	35	0.08
37		Rootlets	6 000	130	15	80	35	–

*In certain cases, a range of values is given in preference to a mean figure, when mean values obtained during measurements differ by more than 300%.

Table 94 *(cont.)*

Code		Raw material	Selenium Se*	Iodine I	Molybdenum Mo	Nickel Ni	Fluorine F
	I	**Major cereals**					
1		Oats	0.1	0.1	0.6	0.8	1.2
2		Wheat — soft	0.05–0.2	0.04	0.3	3	2
3		Maize (corn, USA)	0.05–0.3	0.05	0.25	1.5	0.8
4		Barley — 2 row	0.03–0.3	0.04	0.35	0.3	2.8
5		Barley — 6 row	0.03–0.3	0.04	0.35	0.3	2.8
6		Sorghum — low tannin	0.07	0.02	0.4	1	0.7
7		Sorghum — high tannin	0.07	0.02	0.4	1	0.7
	II	**Minor cereals**					
8		Oats — naked	–	–	–	–	–
9		Oats — dehulled	–	–	–	–	–
10		Oats — flaked	–	–	0.2	–	–
11		Wheat — hard	0.06	–	–	–	–
12		Millet	0.7	–	–	–	–
13		Barley — naked	–	–	–	–	–
14		Rice	0.1	–	0.5	0.2	–
15		Buckwheat	–	–	–	–	–
16		Rye — winter	0.03–0.15	0.07	0.18	1.3	–
17		Triticale — French	–	–	–	–	–
	III	**Cereal by-products**					
		From hard wheat					
18		Middlings	–	–	–	–	–
19		Shorts	–	–	–	–	–
20		Bran	–	–	–	–	–
		From soft wheat					
21		Middlings	–	–	–	–	–
22		Germ	0.8	–	–	–	0.9
23		White shorts	0.7	0.11	0.07	–	–
24		Red shorts	0.7	0.11	0.07	–	–
25		Fine bran	0.05–1.2	0.08	0.55	1.2	3.5
26		Coarse bran	0.05–1.2	0.08	0.55	1.2	3.5
		From maize (corn, USA)					
27		Distiller's dried grains and solubles (DDGS)	0.3	0.03	–	–	–
28		Germ	0.2	0.8	1.3	5	4
29		Gluten feed	0.1–10	–	–	–	–
30		Gluten meal 40	0.1–10	–	–	–	–
31		Gluten meal 60	0.07	–	–	–	–
32		Cobs	–	–	–	–	–
33		Bran	0.2	–	–	–	–
34		Germ — oilmeal (expeller)	0.2	–	–	–	1
35		Germ — oilmeal (solvent)					
		From barley					
36		Brewer's grains	0.06–0.7	–	0.5	–	–
37		Rootlets	0.6	–	0.5	0.2	–

*Two groups of values may be found for selenium, with those from France tending to be lower.

Table 94 *(cont.)*

Code	Raw material	Sulphur S	Iron Fe	Copper Cu	Zinc Zn	Manganese Mn	Cobalt Co
	III Cereal by-products *(cont.)*						
	From rice						
38	Broken	–	–	–	–	–	–
39	Shorts	1 600	170	8	30	300	–
40	Bran	1 800	190	15	30	130–400	0.1
	IV Ensiled cereals						
41	Ears and stalk tops						
42	Ears with husks						
43	Ears without husks						
44	Grains						
45	Whole plants						
	V Carbohydrate raw materials						
46	Maize starch	–	–	–	–	–	–
47							
48	Banana	–	4	1	1	1	–
49							
50	Carob fruit	500	140	3	20	7	–
51	Carob germ	4 500	125	–	50	65	–
52	Chestnut	300	9	2	–	6	–
53							
54	Breadfruit	–	–	–	–	–	–
55	Acorns	400	–	–	–	–	–
56							
57	Dasheen	–	–	–	–	–	–
58	Sugar beet molasses	4 000	50–150	5–18	25	5–20	0.5
59	Sugar cane molasses	3 500	250	1–60	15	45	0.9
60	Apples	80	2–10	1	1	2	0.1
	VI Roots and tubers						
67	Sugar beet	200	10–100	1	6	20	0.02
68	Fodder sugar beet	200	10	1	5	8	0.01
69	Fodder beet	200	4–20	1	4	4	0.01
70	Sugar beet pulp	3 500	450	5–15	1–30	20–80	0.1
71	Arrowroot	–	–	–	–	–	–
72	Carrot	200	3–20	1	4	2	0.02
73	Chicory leaves	–	100	2	8	30	–
74	Chicory roots	–	–	–	–	–	–
75	Yam	–	–	–	–	–	–
76	Cassava	5 000	11	4	15	30	0.04
77							
78	Turnip	–	3	1	–	5	–
79	Sweet potato	400	15	2	–	4	0.005
80	Potato starch	–	10	–	–	–	–
81	Potato—raw	300	15	2	4	2	0.01
82	Potato—dried	1 100	55	7	15	7	0.03

Table 94 *(cont.)*

Code		Raw material	Selenium Se	Iodine I	Molybdenum Mo	Nickel Ni	Fluorine F
	III	**Cereal by-products** *(cont.)*					
		From rice					
38		Broken	–	–	–	–	–
39		Shorts	–	–	–	–	–
40		Bran	0.15	–	0.3	–	–
	IV	**Ensiled cereals**					
41		Ears and stalk tops					
42		Ears with husks					
43		Ears without husks					
44		Grains					
45		Whole plants					
	V	**Carbohydrate raw materials**					
46		Maize starch	–	–	–	–	–
47 48 49		Banana	–	0.02	–	0.2	0.1
50		Carob fruit	–	–	–	–	–
51		Carob germ	–	–	–	–	–
52 53		Chestnut	–	–	–	–	–
54		Breadfruit	–	–	–	–	–
55 56		Acorns	–	–	–	–	–
57		Dasheen	–	–	–	–	–
58		Sugar beet molasses	–	1.5	0.2	4.5	–
59		Sugar cane molasses	–	–	–	–	–
60		Apples	–	0.02	–	0.1	0.06
	VI	**Roots and tubers**					
67		Sugar beet	–	–	0.02	–	–
68		Fodder sugar beet	–	–	0.03	–	–
69		Fodder beet	–	–	0.01	–	–
70		Sugar beet pulp	0.03	1.8	0.3	2	6
71		Arrowroot	–	–	–	–	–
72		Carrot	–	0.1	0.05	–	0.2
73		Chicory leaves	–	–	–	–	–
74		Chicory roots	–	–	–	–	–
75		Yam	–	–	–	–	–
76 77		Cassava	0.1	–	0.05	0.9	4.5
78		Turnip	–	0.05	–	–	–
79		Sweet potato	–	0.02	–	–	–
80		Potato starch	–	–	–	–	–
81		Potato—raw	0.02	0.05	0.05	0.2	1.2
82		Potato—dried	0.06	0.2	0.2	–	3

Table 94 *(cont.)*

Code		Raw material	Sulphur S	Iron Fe	Copper Cu	Zinc Zn	Manganese Mn	Cobalt Co
	VI	**Roots and tubers** *(cont.)*						
83		Potato starch flour	–	–	–	–	–	–
84		Potato flakes	1 100	55	2	16	6	0.01
85		Potato proteins	2 300	50	10	25	5	–
86		Potato pulp	600	300	15	10	30	–
87		Swedes	–	–	–	–	–	–
88		Jerusalem artichokes	200	30	1	–	2	–
	VII	**Industrial by-products**						
89		Citrus pulp	600	150	6	12	6	–
90		Coffee grounds	–	–	–	–	–	–
91		Apple pomace	1 900	300	15	10	10	–
92		Grape pips	–	–	–	–	–	–
93		Grape pulp	4 000	80	20	20	–	–
94		Grapeseed oilmeal	1 900	170	170	15	15	–
95		Tomato skins	–	–	–	–	–	–
96		Tomato pips	–	–	–	–	–	–
97		Tomato pulp	–	–	–	–	–	–
98		Vinasse	–	–	–	–	–	–
99 / 100		Vinasse waste	12 000	200	10	30	25	–
	VIII	**Various vegetable products**						
101		Cocoa hulls	–	500	35	50	70	–
102		Rapeseed hulls	–	–	–	–	–	–
103/106		Cabbages	680	8	1	3	3	–
107		Cereal straw ⎫	–	230	6	25	70	–
108		Grassmeal ⎭						
109		Dehydrated Lucerne meal 21	4 000	280	8	20	40	0.1
110		Dehydrated Lucerne meal 17	3 200	270	8	20	40	0.1
111		Dehydrated Lucerne meal 15	2 000	230	8	20	30	0.1
112		Dehydrated Lucerne meal 12	1 500	230	7	20	30	0.1
113		Lucerne protein concentrate	–	–	–	–	–	–
114		Soya-bean hulls	1 200	2 900	15	60	15–50	–
	IX	**Vegetable proteins**						
115		Full fat rapeseed	–	–	–	–	–	–
116		Field beans	2 600	75	10	45	15	0.35
117		Haricot beans	2 000	70	7	4–50	15–180	0.30
118		Lentils	2 500	80	10	30	15	0.3
119		Sweet white lupins*	2 400	55	6	35	800–3000	0.04
120 / 121		Peas	2 000	75	7	40	10	0.07
122		Full fat soya-beans	3 000	90	15	40	25	–
123		Soya protein isolate	–	–	–	–	–	–

*Manganese content of lupins, which is always high, varies with variety, e.g. Lucky 800, Kalina 1600, Lublanc 3000.

Table 94 (cont.)

Code		Raw material	Selenium Se	Iodine I	Molybdenum Mo	Nickel Ni	Fluorine F
	VI	**Roots and tubers** (cont.)					
83		Potato starch flour	–	–	–	–	–
84		Potato flakes	0.06	0.2	0.2	–	2
85		Potato proteins	–	–	–	–	–
86		Potato pulp	–	–	–	–	–
87		Swedes	–	–	–	–	–
88		Jerusalem artichokes	–	–	–	–	–
	VII	**Industrial by-products**					
89		Citrus pulp	–	–	–	–	–
90		Coffee grounds	–	–	–	–	–
91		Apple pomace	–	–	–	–	–
92		Grape pips	–	–	–	–	–
93		Grape pulp	–	–	–	–	–
94		Grapeseed oilmeal	–	–	–	–	–
95		Tomato skins	–	–	–	–	–
96		Tomato pips	–	–	–	–	–
97		Tomato pulp	–	–	–	–	–
98		Vinasse	–	–	–	–	–
99/100		Vinasse waste	–	–	–	–	–
	VIII	**Various vegetable products**					
101		Cocoa hulls	–	–	–	–	–
102		Rapeseed hulls	–	–	–	–	–
103/106		Cabbages	0.012	0.06	–	0.1	0.05
107		Cereal straw }	–	–	–	–	–
108		Grassmeal }					
109		Dehydrated Lucerne meal 21	0.03–0.3	–	0.6	–	–
110		Dehydrated Lucerne meal 17	0.03–0.3	–	0.6	1.5	10
111		Dehydrated Lucerne meal 15	0.03–0.3	–	0.6	–	–
112		Dehydrated Lucerne meal 12	0.03–0.3	–	0.5	1.5	10
113		Lucerne protein concentrate	–	–	–	–	–
114		Soya-bean hulls	–	–	–	–	–
	IX	**Vegetable proteins**					
115		Full fat rapeseed	–	–	–	–	–
116		Field beans	0.02	–	0.7	0.12	1
117		Haricot beans	0.2	–	0.7	3	1
118		Lentils	0.10	–	–	3	0.8
119		Sweet white lupins	0.07	–	3	5	–
120/121		Peas	0.18	0.15	1	3	1.4
122		Full fat soya-beans	0.5	0.05	2.5	6	–
123		Soya protein isolate	–	–	–	–	–

Table 94 (cont.)

Code	Raw material	Sulphur S	Iron Fe	Copper Cu	Zinc Zn	Manganese Mn	Cobalt Co
X	**Oil meals**						
124	Peanut meal	3 000	20	15	55	40	0.3
125	Dehulled rapeseed meal	–	–	–	–	–	–
126/127	Rapeseed meal — expeller/solvent	2 500	160	5	55	55	0.1
128	Copra meal	3 000	550	20	60	65	0.2
129	Cotton seed meal	3 000	100	20	60	20	0.15
130	Palm kernel meal	3 000	230	25	40	190	0.13
131	Soya-bean meal 44	4 000	170	20	55	30	0.25
132	Soya-bean meal 48	4 500	150	20	50	25–145	0.25
133	Soya-bean meal 50	4 500	150	15	50	35	0.25
134	Sunflower seed meal	4 000	30	25	80	30	0.15
XI	**Microbial proteins/algae**						
135/137	Algae	20 000	200–1000	2–15	50–200	10–1200	1–10
138	Brewer's yeast	4 500	50–300	25	60	5	0.2
139	Distiller's yeast	4 500	250	65	10	20	0.1
140	Torula yeast	4 000	100	15	90	13	1.6
141	Lactic yeast	–	–	–	–	–	–
142	Fodder protein	–	–	–	–	–	–
143	ICI Pruteen	10 000	300	20	40	–	–
XII	**Animal by-products**						
144	Fish protein concentrate	–	–	–	–	–	–
145/146	Krill meal	–	320	210	45	–	–
147	Fish meal 60% crude protein	5 000	350	8	150	10–35	0.15
148/149	Fish meal 65% crude protein	6 000	300	8	110	10	0.1
150/151	Fish meal 72% crude protein	6 000	250	5	100	10	0.1
152	Fish solubles	3 000	300	40	40	15	0.3
153	Greaves	6 000	900	30	15	20	–
154	Feather meal	9 000	70	–	70	7	–
155	Blood meal	4 000	2 000	9	20–300	6	0.1
156	Meat meal 50% crude protein ⎫	4 500	500	10	100	5–20	0.03
157	Meat meal 55% crude protein ⎭						
158	Meat meal defatted 50% crude protein	–	–	–	–	–	–
159	Meat meal defatted 55% crude protein	–	–	–	–	–	–
160	Meat meal defatted 60% crude protein	–	–	–	–	–	–
161	Meat and bone meal	–	–	–	–	–	–
162	Poultry litter	–	–	–	–	–	–
163	Poultry offal meal	–	–	–	75	5	–
174	Poultry hatchery waste	–	–	–	–	–	–
XIII	**Dairy products**						
	Liquid						
165	Milk	250	0.2–0.8	0.05–0.2	4	0.04	0.001
166	Skimmed milk	350	0.5–2	0.04–0.2	4	0.07–0.2	0.001
167	Sweet buttermilk	100	1	–	4	0.2	0.001
168/169	Separated whey	–	1	0.4	0.5	0.05–0.5	–
170	Mixed separated whey	–	–	–	–	–	–
171	Acid goat whey	–	–	–	–	–	–
172/173	Whey casein	–	–	–	–	–	–

Table 94 (cont.)

Code	Raw material	Selenium Se	Iodine I	Molybdenum Mo	Nickel Ni	Fluorine F
	X Oil meals					
124	Peanut meal	0.1	0.4	1.5	10	–
125	Dehulled rapeseed meal	–	–	–	–	–
126/127	Rapeseed meal — expeller/solvent	0.07–0.6	0.6	0.45	2	14
128	Copra seed meal	–	1.3	0.6	5	7
129	Cotton meal	0.7	0.1	0.8	3	3
130	Palm kernel meal	0.12	1.1	0.4	–	–
131	Soya-bean meal 44	0.1–0.5	0.15	2	19	7
132	Soya-bean meal 48	0.1–0.5	0.15	2	19	7
133	Soya-bean meal 50	0.1–0.5	0.15	2	19	7
134	Sunflower seed meal	0.5	0.6	0.8	5	2.5
	XI Microbial proteins/algae					
135/137	Algae	0.5	250–4 000	1–15	–	320
138	Brewer's yeast	0.1–1	0.01	1	4.8	2.5
139	Distiller's yeast	0.1	0.01	0.35	–	–
140	Torula yeast	0.06	–	0.1	1.6	7
141	Lactic yeast	–	–	–	–	–
142	Fodder protein	–	–	–	–	–
143	ICI Pruteen	0.4	–	–	–	8
	XII Animal by-products					
144	Fish protein concentrate	–	–	–	–	–
145/146	Krill meal	2	–	–	–	11
147	Fish meal 60% crude protein	1–5	2	0.1	1	150
148/150	Fish meal 65% crude protein	1–4	2	0.1	1	–
149/151	Fish meal 72% crude protein	1–4	5	0.1	1	–
152	Fish solubles	2	1	–	–	–
153	Greaves	–	–	–	–	–
154	Feather meal	0.3	–	–	–	–
155	Blood meal	0.2–1.6	0.8	0.2	0.6	12
156	Meat meal 50% crude protein	0.3	1.3	0.4	2	60
157	Meat meal 55% crude protein					
158	Meat meal defatted 50% crude protein	–	–	–	–	–
159	Meat meal defatted 55% crude protein	–	–	–	–	–
160	Meat meal defatted 60% crude protein	–	–	–	–	–
161	Meat and bone meal	–	–	–	–	–
162	Poultry litter	–	–	–	–	–
163	Poultry offal meal	1	–	–	–	–
164	Poultry hatchery waste	–	–	–	–	–
	XIII Dairy products					
	Liquid					
165	Milk	0.03	0.02–0.15	0.02–0.1	0.03	0.03–0.15
166	Skimmed milk	0.05	0.03	0.02	–	0.15
167	Sweet buttermilk	–	0.02	–	–	0.1
168/169	Separated whey	–	–	–	0.04	–
170	Mixed separated whey	–	–	–	–	–
171	Acid goat whey	–	–	–	–	–
172/173	Whey casein	–	–	–	–	–

Table 94 *(cont.)*

Code	Raw material	Sulphur S	Iron Fe	Copper Cu	Zinc Zn	Manganese Mn	Cobalt Co
	XIII Dairy products *(cont.)*						
	Dried						
174	Milk	2 300	3–10	0.6–2	10–50	0.2–1	0.01
175	Skimmed milk	3 200	3–10	2.5	25–70	0.7–3	0.01
176	Sweet buttermilk	3 500	3–12	2	35	2	0.01
177/178	Casein	6 000	10	0.2	30	0.4	–
179	Sodium caseinate	–	–	–	–	–	–
180/181	Separated whey	6 000	6–15	1–6	3–20	0.4–5	0.1
182	Sweet ewe whey	1 300	6	1	2	0.3	–
183	Milk permeate	–	10	2	30	0.5	–
184	Mixed whey permeate	–	–	–	–	–	–
185	Partially delactosed whey	3 500	20	1	12	4	–
186	Whey protein	–	–	–	–	–	–

Code	Raw material	Selenium Se	Iodine I	Molybdenum Mo	Nickel Ni	Fluorine F
	XIII Dairy products *(cont.)*					
	Dried					
174	Milk	0.3	0.3–1	0.3	0.2	1
175	Skimmed milk	0.2	1	0.2	–	1
176	Sweet buttermilk	0.1	0.2	–	–	1
177/178	Casein	0.15	–	–	–	0.2
179	Sodium caseinate	–	–	–	–	–
180/181	Separated whey	0.06–0.6	–	–	0.8	0.4
182	Sweet ewe whey	–	–	–	–	–
183	Milk permeate	–	–	–	–	–
184	Mixed whey permeate	–	–	–	–	–
185	Partially delactosed whey	0.06	–	–	–	–
186	Whey protein	–	–	–	–	–

Table 95 Mineral sources — pure products (reagent grade). For natural and feed grade products see Table 93, numbers 190–224

Principal element	Salt	Formula	Molecular weight (g/mol)	Composition (%)			
Calcium	Calcium carbonate	$CaCO_3$	100.0	Ca	= 40.0	CO_3	= 60.0
	Calcium chloride — anhydrous	$CaCl_2$	111.0	Ca	= 36.0	Cl	= 64.0
	Calcium sulphate	$CaSO_4$	136.1	Ca	= 29.4	SO_4	= 70.6
Chlorine	Sodium chloride	NaCl	58.4	Cl	= 60.6	Na	= 39.4
	Potassium chloride	KCl	74.6	Cl	= 47.6	K	= 52.4
	Calcium chloride — anhydrous	$CaCl_2$	111.0	Cl	= 64.0	Ca	= 36.0
Cobalt	Cobalt carbonate	$CoCO_3$	118.9	Co	= 49.6	CO_3	= 50.4
	Cobalt chloride hexahydrate	$CoCl_2 \cdot 6H_2O$	237.9	Co	= 24.9	Cl	= 29.8
	Cobalt sulphate heptahydrate	$CoSO_4 \cdot 7H_2O$	281.1	Co	= 21.0	SO_4	= 34.2
Copper	Copper chloride — anhydrous	$CuCl_2$	134.4	Cu	= 47.3	Cl	= 52.7
	Copper oxide	CuO	79.5	Cu	= 79.9	–	
	Copper sulphate — anhydrous	$CuSO_4$	159.6	Cu	= 39.8	SO_4	= 60.2
	Copper sulphate pentahydrate	$CuSO_4 \cdot 5H_2O$	249.7	Cu	= 25.5	SO_4	= 38.5
Iron	Iron chloride tetrahydrate	$FeCl_2 \cdot 4H_2O$	198.8	Fe	= 28.1	Cl	= 35.7
	Ammonium ferric citrate[a]	–	–	Fe	= 17.5	–	
	Iron sulphate heptahydrate	$FeSO_4 \cdot 7H_2O$	278.0	Fe	= 20.1	SO_4	= 34.6
Fluorine	Potassium fluoride — anhydrous	KF	58.1	F	= 32.7	K	= 67.3
	Potassium fluoride dihydrate	$KF \cdot 2H_2O$	94.1	F	= 20.2	K	= 41.5
	Sodium fluoride	NaF	42.0	F	= 45.2	Na	= 54.8
Iodine	Calcium iodate	$Ca(IO_3)_2$	390.0	I	= 65.1	Ca	= 10.3
	Potassium iodate	KIO_3	214.0	I	= 59.3	K	= 18.3
	Potassium iodide	KI	166.0	I	= 76.4	K	= 23.6
	Sodium iodide	NaI	150.0	I	= 84.7	Na	= 15.3
	Pentacalcium orthoperiodate	$Ca_5(IO_6)_2$	646.2	I	= 19.6	Ca	= 30.9
Magnesium	Magnesium chloride hexahydrate	$MgCl_2 \cdot 6H_2O$	203.3	Mg	= 12.0	Cl	= 34.9
	Magnesium hydroxycarbonate pentahydrate	$(MgCO_3)_4Mg(OH)_2 \cdot 5H_2O$	485.5	Mg	= 25.0	CO_3	= 50.0
	Magnesium oxide	MgO	40.3	Mg	= 60.3		
	Magnesium sulphate — anhydrous	$MgSO_4$	120.4	Mg	= 20.2	SO_4	= 79.8
	Magnesium sulphate heptahydrate	$MgSO_4 \cdot 7H_2O$	246.5	Mg	= 9.9	SO_4	= 39.0
Manganese	Manganese carbonate	$MnCO_3$	114.9	Mn	= 47.8	CO_3	= 52.2
	Manganese chloride dihydrate	$MnCl_2 \cdot 2H_2O$	161.9	Mn	= 33.9	Cl	= 43.8
	Manganese chloride tetrahydrate	$MnCl_2 \cdot 4H_2O$	197.9	Mn	= 27.8	Cl	= 35.9
	Manganese oxide	MnO	70.9	Mn	= 77.4	–	
	Manganese sulphate monohydrate	$MnSO_4 \cdot H_2O$	169.1	Mn	= 32.5	SO_4	= 56.9
	Manganese sulphate tetrahydrate	$MnSO_4 \cdot 4H_2O$	223.1	Mn	= 24.6	SO_4	= 43.1
Molybdenum	Sodium molybdate dihydrate	$Na_2MoO_4 \cdot 2H_2O$	241.9	Mo	= 39.7	Na	= 19.0
	Molybdenum oxide	MoO_3	143.9	Mo	= 66.7	–	–
Phosphorus	Orthophosphoric acid	H_3PO_4	98.0	P	= 36.1	–	
	Monocalcium phosphate monohydrate [b]	$Ca(H_2PO_4)_2 \cdot H_2O$	252.1	P	= 24.6	Ca	= 15.9
	Calcium phosphate — anhydrous	$CaHPO_4$	136.1	P	= 22.8	Ca	= 29.5
	Calcium phosphate dihydrate [c]	$CaHPO_4 \cdot 2H_2O$	172.1	P	= 18.0	Ca	= 23.3
	Tricalcium phosphate	$Ca_3(PO_4)_2$	310.2	P	= 20.0	Ca	= 38.8
	Monopotassium phosphate [d]	KH_2PO_4	136.1	P	= 22.8	K	= 28.7
	Dipotassium phosphate [e]	K_2HPO_4	174.2	P	= 17.8	K	= 44.9
	Monosodium phosphate monohydrate [f]	$NaH_2PO_4 \cdot H_2O$	138.0	P	= 22.4	Na	= 16.7
	Monosodium phosphate dihydrate [f]	$NaH_2PO_4 \cdot 2H_2O$	156.0	P	= 19.9	Na	= 14.7
	Disodium phosphate [g]	Na_2HPO_4	142.0	P	= 21.8	Na	= 32.4
Potassium	Potassium acetate	KCH_3COO	98.1	K	= 39.8	Acetate	= 60.2
	Potassium bicarbonate [h]	$KHCO_3$	100.1	K	= 39.1	HCO_3	= 60.9
	Potassium carbonate	K_2CO_3	138.2	K	= 56.6	CO_3	= 43.4
	Potassium chloride	KCl	74.6	K	= 52.4	Cl	= 47.6
	Tripotassium phosphate	K_3PO_4	212.3	K	= 55.2	P	= 14.6
	Potassium sulphate	K_2SO_4	174.3	K	= 44.9	SO_4	= 55.1
Selenium	Selenium selenite	Na_2SeO_3	172.9	Se	= 45.6	Na	= 26.6
	Selenium selenate	Na_2SeO_4	189.0	Se	= 41.8	Na	= 24.3
Sodium	Sodium acetate	$NaCH_3COO$	82.0	Na	= 28.0	Acetate	= 72.0
	Sodium bicarbonate [i]	$NaHCO_3$	84.0	Na	= 27.4	HCO_3	= 72.6
	Sodium carbonate	Na_2CO_3	106.0	Na	= 43.4	CO_3	= 56.6
	Sodium chloride	NaCl	58.4	Na	= 39.4	Cl	= 60.6
	Sodium sulphate — anhydrous	Na_2SO_4	142.0	Na	= 32.4	SO_4	= 67.6
Zinc	Zinc carbonate	$ZnCO_3$	125.4	Zn	= 52.1	CO_3	= 47.9
	Zinc chloride	$ZnCl_2$	136.3	Zn	= 48.0	Cl	= 52.0
	Zinc oxide	ZnO	81.4	Zn	= 80.3	–	
	Zinc sulphate monohydrate	$ZnSO_4 \cdot H_2O$	179.4	Zn	= 36.4	SO_4	= 53.5
	Zinc sulphate heptahydrate	$ZnSO_4 \cdot 7H_2O$	287.5	Zn	= 22.8	SO_4	= 33.4

Official names of these salts are as follows
[a] Ammonium Fe(III) citrate.
[b] Calcium dihydrogenphosphate.
[c] Calcium hydrogenphosphate.
[d] Potassium dihydrogenphosphate.
[e] Dipotassium hydrogenphosphate.
[f] Sodium dihydrogenphosphate.
[g] Disodium hydrogenphosphate.
[h] Potassium hydrogencarbonate.
[i] Sodium hydrogencarbonate.

Mono, bi and tri calcium phosphates may equally be called mono, di and tri basic, respectively.

Table 96 Relative values of various minerals and their oxides

Al/Al_2O_3	=	0.53	Al_2O_3/Al	= 1.89
Ca/CaO	=	0.71	CaO/Ca	= 1.40
Fe/Fe_2O_3	=	0.70	Fe_2O_3/Fe	= 1.43
Mg/MgO	=	0.60	MgO/Mg	= 1.66
Mn/MnO	=	0.77	MnO/Mn	= 1.29
Na/NaO	=	0.59	NaO/Na	= 1.70
P/P_2O_5	=	0.44	P_2O_5/P	= 2.29
P/H_3PO_4	=	0.32	H_3PO_4/P	= 3.16
H_3PO_4/P_2O_5	=	1.38	P_2O_5/H_3PO_4	= 0.72
Si/SiO_2	=	0.47	SiO_2/Si	= 2.14

Table 97 Vitamin composition (mg/kg) of common raw materials and those considered to be useful sources

Code	Raw material	E	B_1	B_2	Calcium pantothenate	B_6	B_{12}	Niacin	Folic acid	Biotin	Choline
1	Oats	18	6	1.5	13	2.0	0	15	0.35	0.25	1 000
2	Wheat	15	4.5	1.2	12	3.5	0	55	0.30	0.10	800
3	Maize	20	3.7	1.2	6	5.0	0	22	0.30	0.07	500
4–5	Barley	25	4.5	1.3	7	3.0	0	55	0.40	0.15	1 000
6–7	Sorghum	13	4.0	1.3	12	4.0	0	40	0.20	0.20	600
22	Wheatgerm	150	22.0	5.0	18	12.0	0	55	2.50	0.25	3 100
23–24	Shorts	30	18	3.0	22	8.0	0	95	1.40	0.15	1 100
25	Fine wheat bran	17	8.0	3.0	30	12	0	220	1.30	0.15	1 000
27	Distiller's dried grains and solubles — maize	40	3.5	9.0	14	8	0.013	80	0.85	0.40	2 000
28	Maize germ	90	10.0	3.7	4	6	0	42	0.45	0.20	1 900
31	Maize gluten meal 60	26	0.3	2.1	12	10	0	65	0.23	0.20	400
36	Brewer's grains	30	0.7	1.0	10	1	0	43	0.22	0.08	1 550
58	Sugar beet molasses	2	1.0	2.0	5	5	0	48	0.21	0.05	1 070
62	Animal fat	8	–	–	–	–	–	–	–	–	–
63	Poultry fat	19	–	–	–	–	–	–	–	–	–
64	Vegetable oil	60	–	–	–	–	–	–	–	–	–
65	Lard	22	–	–	–	–	–	–	–	–	–
66	Tallow	25	–	–	–	–	–	–	–	–	–
70	Sugar beet pulp	–	0.3	1.0	1	2	0	19	–	–	800
76	Cassava	–	1.6	0.8	1	1	0	3	–	–	–
82	Potatoes — dried	–	1.0	0.6	2	2.6	0	10	0.80	0.10	825
98	Vinasse	–	81	28.5	–	–	0	269	–	–	–
108	Grassmeal	111	4.8	10.0	18	9	0	78	–	0.22	1 470
110	Dehydrated Lucerne meal 17	120	3.5	17	30	8	0	46	3.00	0.35	1 600
112	Dehydrated Lucerne meal 12	40	2.8	11	25	7	0	35	2.50	0.30	1 500
116	Field beans	9	5.5	2.6	3	–	0	25	–	0.09	1 670
120	Peas	8	1.6	1.0	5	10	0	17	–	0.20	650
122	Full fat soya-beans	55	10.0	2.6	16	10	0	23	3.52	0.30	2 000
124	Peanut meal	3	7.5	6.0	50	6	0	170	0.36	0.90	1 700
125/126	Rapeseed meal	10	1.8	3.5	9	14	0	159	–	0.94	6 500
129	Cotton seed meal	14	7.5	4.5	13	6	0	39	2.20	0.50	2 800
131	Soya-bean meal 44	3.5	3.0	3.0	15	7	0	40	3.60	0.35	2 800
133	Soya-bean meal 50	4.0	2.4	3.0	14	8	0	22	3.60	0.35	2 750
134	Sunflower seed meal	12	33.9	3.2	10	13	0	200	0.42	1.40	2 500
135/137	Algae	108	2.7	7.5	29	1	0.012	29	1.91	0.40	–
138	Brewer's yeast	2	85	45	110	30	0.008	450	11.20	1.10	2 800
139	Distiller's yeast	0	40	35	85	–	–	300	11.00	0.27	3 400
143	ICI Pruteen	25	5	37	10	2	0.030	52	14.00	2.90	10
150	Fishmeal 65% crude protein	6	0.1	7.5	9	4	0.300	150	0.24	0.25	3 080
151	Fishmeal 70% crude protein	17	0.2	9.0	13	5	0.400	120	0.52	0.35	4 550
152	Fish solubles	–	6.8	15	50	12	0.500	210	0.73	0.25	4 000
156	Meat meal 50% crude protein	1	1.1	5.0	4	2	0.090	50	0.60	0.14	2 000
168	Sweet separated whey	0.5	3.7	26	46	3	0.020	11	0.90	0.40	1 980
173	Lactin casein whey	–	0.4	1.5	3	0.5	0.010	1	0.40	0.04	210
175	Skimmed milk powder	8	3.5	20.0	34	4	0.025	12	0.62	0.33	1 400

Table 98 Major sources of linoleic acid and average contents (%)

Code	Raw material	%
1	Oats	2.68
2	Wheat	0.91
3	Maize	1.89
4–5	Barley	0.90
6–7	Sorghum	1.35
21	Middlings	1.29
23	White shorts	1.95
24	Red shorts	1.90
25–26	Wheat bran	1.90
28	Maize germ	9.32
29	Gluten feed	1.34
30	Gluten meal 40	1.28
31	Gluten meal 60	1.23
33	Maize bran	2.85
34	Germ oilmeal — expeller	3.39
35	Germ oilmeal — solvent	0.80
62	Animal fat	8.48
63	Poultry fat	24.70
64	Vegetable (maize) oil	53.20
66	Tallow	2.63
76	Cassava	0.24
115	Full fat rapeseed	5.90
120/121	Peas	0.56
122	Full fat soya-beans	10.46
126	Rapeseed meal — expeller	1.13
127	Rapeseed meal — solvent	0.25
133	Soya-bean meal 50	0.96
134	Sunflower seed meal	1.15
138	Brewer's yeast	0.03
139	Distiller's yeast	0.04
149	Fish meal 72% crude protein	0.23
156	Meat meal 50% crude protein	0.44
158	Meat meal defatted 50% crude protein	0.21
174/175	Milk powder	0.02

Table 99 Sources of xanthophylls

Code	Raw material	Concentration (mg/kg)	Relative efficiency *
1	Oats	0	
2	Wheat	0	
3	Maize	17	0.85
4–5	Barley	0	
6–7	Sorghum	0	
29	Gluten feed	22	0.85
30	Gluten meal 40	140	0.85
31	Gluten meal 60	280	0.85
109	Lucerne meal 21	280	0.72
110	Lucerne meal 17	170	0.72
111	Lucerne meal 15	170	0.72
112	Lucerne meal 12	140	0.72
113	Lucerne concentrate	960	0.72
133	Soya-bean meal 50	0	
135	Algae — Spirulina	3 000	0.75
	Marigold meal	5 400	0.70
	Beta-apo-carotenal	105 000	0.94

*Estimated with reference to the ethyl ester of beta-apo-8-carotenoic acid and the yolk colour thus produced (measured with a Roche fan), with a dietary xanthophyll level of 10 mg/kg.

Table 100 Maximum dietary inclusion level of certain raw materials for pigs

Code	Raw material	Limiting factors or possible adverse effects		*	Recommended maximum inclusion levels **(%)			
		Antinutritive factors	Other factors		Suckling	Weaner	Fattening	Breeding
1	Oats		Fibre		0	5	20	20
2	Wheat				40	–	–	–
3	Maize		Fibre		40	–	–	–
4–5	Barley		Fibre (winter barley)		40	–	–	–
6–7	Sorghum	Tannins			?	?	–	–
16	Rye — winter		(Ergot)		?	–	–	–
17	Triticale — French				?	–	–	–
23–24	Shorts		Fibre of various types		0	10	30	30
25–26	Wheat bran				0	5	15	15
58–59	Molasses	Excess potassium, diarrhoetic effect + level of water		L	–	5	8	20
				F				
				G	5	5	10	10
62–63	Animal fats	Chemical structure (fatty acid composition) + method of manufacture		F		10	10	10
64	Vegetable oil			G	7	5	21	5
67–69	Sugar beet (DM basis)	Level of water, bulkiness			0	0	40	60 (gestation)
70	Dehydrated sugar beet pulp	Diarrhoetic			0	0	10	15
76	Cassava I pelleted	Hydrocyanic acid	Fibre, ash (silica), microbiological quality		0	0	20	20
77	Cassava II roots				20	30	30	20
82	Potatoes — dried				0	0	40	40
85	Potato proteins				10	5	–	–
109/112	Dehydrated Lucerne meal		Fibre, unsaturated fatty acids, palatability		0	0	5	10
116	Field beans — usual	Tannins			0	0	15	10
	Field beans—low tannin				0	0	20	15
119	Sweet lupin		Alkaloids, α-galactosides		0	5	10	0
120	Winter peas	Antitrypsin factors			0	15	20	20
121	Spring peas				0	20	25	25
124	Peanut meal		Aflatoxins,		0	0	10	0
125	Dehulled rapeseed meal 00†				0	0	15	?
127	Rapeseed meal	Glucosinolates	Fibre palatability		0	0	5	?
	Rapeseed meal 00†				0	0	10	?
133	Soya-bean meal 50				20	25	–	–
134	Sunflower seed meal		Fibre		0	0	15	15
138/139	Yeast — all sources		Residual components of substrate culture		0	0	5	10

*Form of diet: G–granule, F–flour, P–pellet, M–meal, L–liquid.
** — No constraint, 0 — use not adivsed, ? — insufficient data.
† 00—double zero variety.

Table 100 (*cont.*)

Code	Raw material	Limiting factors or possible adverse effects		*	Recommended maximum inclusion levels **(%)			
		Antinutritive factors	Other factors		Suckling	Weaner	Fattening	Breeding
140/141	Torula, lactic yeast				10	–	–	–
146	CPSP 80				10	5	–	–
147/151	Fish meals	Source and manufacture, fatty acid composition			10	5	7	7
156/161	Meat meals	Microbial quality, protein digestibility, ash content			0	0	5	5
168/173	Liquid whey	Lactose, water, minerals		L 0		20	40	30 (gestation)
175	Dried skimmed milk	Hygroscopic + method of manufacture		F – G 15		25 10	20 10	10 10
180/181	Dried whey	Lactose, hygroscopic + method of manufacture		F 20 G 15		20 10	20 10	20 10

*Form of diet: G—granule, F—flour, P—pellet, M—meal, L—liquid.
** —No constraint, 0—use not adivsed,?—insufficient data.

Table 101 Maximum dietary inclusion levels of certain raw materials for poultry

Code	Raw material	Limiting factors or possible adverse effects		Recommended maximum inclusion levels (%)	
		Antinutritive factors	Other factors	Young birds	Adults
1	Oats	Antienzymes	β-Glucans		30
2	Wheat			40	None
4–5	Barley	Antienzymes, tannins	β-Glucans	30*	50
6–7	Sorghum	Tannins	If tannin levels > 0.3%	20	40
16	Rye	Polyphenols	β-Glucans	15	25
58–59	Molasses		Excess of potassium	20	
76–77	Cassava	Hydrocyanic acid, Antiphosphatase		15	30
116	Field beans	Tannins	α-Galactosides	30	15
119	Sweet lupin		α-Galactosides, Alkaloids	20	10
120/121	Peas		α-Galactosides	25	20
124	Peanut meal		Mycotoxins?		0
127	Rapeseed meal	Glucosinolates	Sinapine	5*	0
129	Cotton seed meal		Gossypol	8	10
147/151	Fish meals		Unsaturated fatty acids	8	5
156/161	Meat meals	Antibiotin	Excess of calcium	8	0
174/175	Milk powder		Lactose	10	12
					10

*Fattening

Table 102 Maximum permissible levels of certain non-nutritive factors (mg/kg)

Substance	Compound feed		Various raw materials	
Aflatoxin B$_1$	For pigs and poultry	0.02	All	0.05
	For piglets and chicks	0.01	All	10
Castor oil plant — Ricinus communis L. (expressed in terms of castor oil plant husks)	All	10	All	10
Crotalaria L. spp.			All unmilled material	100
Free gossypol	Laying hens and piglets	20	Cotton seed cake or meal	1200
	Other poultry	100	Others	20
	Other pigs and rabbits	60		
Hydrocyanic acid	Chicks	10	Linseed	250
	Others	50	Linseed cake material	350
			Manioc products and almond cakes	100
			Others	50
Rye ergot — Claviceps purpurea (Fr.) Tul			All feeding stuffs containing unground cereals	1000

Index of feed ingredients

The figures shown are the numbers used in Tables 93, 94, 97, 98, 99, 100 and 101

A

Achira (or Queensland arrowroot), 71
Acorns
  dehulled, 55
  whole, 56
Alfalfa (or lucerne)
  meal, 109–112
  protein concentrate (PX1), 113
Algae
  Chlorella, 136
  Scenedesmus, 137
  Spirulina, 135
Algomarine (sea shells), 195
Amino acids
  l-lysine, HCl, 188
  dl-methionine, 187
  methionine hydroxyanalogue (MHA), 189
Animal fat, 62
Apple
  pomace, 91
  whole, 60
Arrowroot, (Queensland- or Achira), 71
Ashes (from hatchery), 199

B

Banana
  ripe, whole, 49
  unripe, silage, 47
  unripe, whole, 48
Barley
  by-products
    brewer's grains, 36
    rootlets, 37
  naked, 13
  six rowed, 5
  two rowed, 4
Bean, common (toasted), 117
Beet
  fodder-, 69
  fodder-sugar, 68
  molasses, 58
  sugar, 67
  sugar, pulp, 70
Bloodmeal, 155
Bone meal
  degelatinized, 220
  raw, 219
Bran
  maize, 33
  hard wheat, 20
  rice, 40
  soft wheat
    coarse, 26
    fine, 25
Bread-fruit, 54
Brewer's grains, 36
Brewer's yeast, 138
Broken rice, 38
Brussels sprouts, 104
Buckwheat, 15
Buttermilk (sweet)
  dried, 176
  liquid, 167
By-products: *see* under different products

C

Cabbages, 103
Calcium
  carbonate
    from quarry
      calcium carbonate, 190
      natural limestone, 191
    from sugar industry, 193
    hatchery ashes, 199
    sea shells, 192
  sources of
    carbonate, 190–193

Calcium
  sources of (cont.)
    egg shells, from egg-breaking plants
      cleaned and dried, 197
      dried, 196
    hatchery ashes, 199
    marl, 198
    natural gypsum, 201
    oyster shells, 194
    Portland cement, 200
    sea shells (Algomarine), 195
Carbohydrate souces, 46–61
Carob (or St John's bread)
  fruit, 50
  germ, 51
Carrots, 72
Casein (dried)
  hydrochloric-, 178
  lactic-, 177
  sodium caseinate, 179
Cassava
  pelleted, 76
  roots, 77
Cauliflowers, 105
Cement (Portland), 200
Cereals
  by-products of, 18–40
  ensiled, 41–45
  minor, 8–17
  major, 1–7
Cereal straw, 107
Chestnut
  dehulled, 52
  whole, 53
Chicory (common)
  leaves, 73
  roots, 74
Chorella (alga), 136
Chloride
  sea salt, 223
  sodium chloride, 222
Citrus fruits (dried pulp), 89
Cobs (of maize), 32
Cocoa hulls, 101
Coffee-grounds, 90
Copra meal, 128
Cotton seed meal, 129

D
DDGS (= maize distiller's dried grains and solubles), 27
Dasheen, 57
Distiller's by-products,
  DDGS, 27
  yeast, 139

E
Ears of maize (ensiled)
  in husks, 42
  with stalk tops, 41
  without husks, 43
Egg shells from egg-breaking plants
  cleaned and dried, 197
  dried, 196
Ewe sweet whey, 182

F
Fats
  animal fat FGAF, 62
  lard, 65
  oil (vegetable), 64
  poultry fat, 63
  tallow, 66
Feather meal, 154
Field beans, 116
Fish
  concentrate (CPSP), 144
  meal
    defatted 65, 150
    defatted 72, 151
    fat 60, 147
    fat 65, 148
    fat 72, 149
  solubles, 152
Flakes
  oats, 10
  potato, 84
Fodder beet, 69
Fodder sugar beet, 68
Fruits
  acorn
    dehulled, 55
    whole, 56
  apple
    pomace, 91
    whole, 60
  banana
    ripe, whole, 49
    unripe, silage, 47
    unripe, whole, 48
  bread-fruit, 54
  carob (or St John's bread), 50
  chestnut
    dehulled, 52
    whole, 53
  dasheen, 57

G
Germ
  carob (or St John's bread), 51
  maize, 28
  oilmeal, maize (expeller), 34
  oilmeal, maize (solvent), 35
  soft wheat, 22

Index

Gluten (maize)
  feed, 29
  meal 40, 30
  meal 60, 31
Goat (acid whey of), 171
Grape
  pips, 92
  pulps, 93
Grapeseed oilmeal, 94
Grass meal, 108
Greaves, 153
Gypsum (natural), 201

H
Hardwheat, 11
  by-products
    bran, 20
    middlings, 18
    shorts, 19
Hatchery
  ashes, 199
  waste, 164
Hulls
  cocoa, 101
  rapeseed, 102
  soya-bean, 114
Hydrochloric casein, 178

J
Jerasulem artichoke (or topinambur),
  fresh, 88

K
Krill
  concentrate, 145
  meal, 146

L
Lactic casein, 177
lactic yeast, 141
Lard, 65
Lentils, 118
Lucerne (or alfalfa)
  meal, 109–112
  protein concentrate (PX1), 113
Lupin, sweet, white, 118
l-Lysine, HC1, 188

M
Maize
  by-products
    bran, 33
    cobs, 32
    distiller's dried grains and solubles
      (DDGS), 27
    germ, 28

    germ, oilcake (expeller), 34
    germ, oilmeal (solvent), 35
    gluten feed, 29
    gluten meal 40, 30
    gluten meal 60, 31
    starch, 46
  grain, 3
  ensiled
    ears and stalk tops, 41
    ears and husks, 42
    ears without husks, 43
    grain, 44
    whole plant, 45
Marl, 198
Marrowstem kale, 106
Meals
  animal origin
    blood, 155
    bone
      degelatinized, 220
      raw, 219
    feather, 154
    fish, 147–151
    greaves, 153
    hatchery waste, 164
    krill, 146
    meat, 156–161
    poultry offal meal, 163
  plant origin
    alfalfa, 109–112
    grass, 108
Meat meal
  defatted 50, 158
  defatted 55, 159
  defatted 60, 160
  fat 50, 156
  fat 55, 157
  with bones, 161
dl-Methionine, 187
Methionine hydroxyanalogue (MHA),
  189
Middlings
  hard wheat, 18
  soft wheat, 21
Milk
  products
    buttermilk, sweet
      dried, 178
      liquid, 167
    casein (dried)
      hydrochloric, 177
      lactic, 178
      sodium caseinate, 179
    permeate, 183
      from mixed whey, 184
    whey, 168–173, 180–182, 184–186
  skimmed
    dried, 175
    liquid, 168
  whey, partially delactosed, dried, 185
  whey, proteins (dried), 186

Milk (*cont.*)
  whole
    dried, 174
    liquid, 165
Millet, 12
Mineral sources, 190–224
Molasses
  beet, 58
  sugar cane, 59

O
Oats, 1
  flakes, 10
  naked, 8
  dehulled, 9
Oil (vegetable-), 64
Oil meals
  copra, 128
  cotton seed, 129
  grape seed, 94
  maize germ
    expeller, 34
    solvent, 35
  palm-kernel, 130
  peanut, 50,124
  rapeseed
    dehulled, 125
    expeller, 126
    solvent extracted, 127
  soya-bean 44, 131
  soya-bean 48, 132
  soya-bean 50, 133
  sunflower seed 34, 134
Oilseeds
  rape: full fat, 115
  soya-bean: full fat, 122
Oyster shells, 194

P
Paddy rice, 14
Palm-kernel meal, 130
Pea
  spring smooth-, 121
  winter smooth-, 120
Peanut meal, 124
Phosphate
  sparingly soluble or insoluble forms
    bone meal
      degelatinized, 220
      untreated, 219
    Ca-Mg-Na, 221
    dicalcium, anhydrous, 214
    dicalcium, hydrated, 213
    mono/dicalcium, 215
    rock, 217
      deflourinated, 218
    tricalcium, 216
  water-soluble forms
    diammonium, 210
    monoammonium, 209
    monocalcium, 212
    dipotassium, 208
    monopotassium, 207
    disodium, anhydrous, 206
    disodium, dihydrate, 205
    monosodium, anhydrous, 204
    monosodium, dihydrate, 203
    sodium tripolyphosphate, 211
Phosphoric acid, 202
Phosphorus sources
  phosphate, 203–221
  phosphoric acid, 202
Pips
  grape, 92
  grapeseed oilmeal, 94
  tomato, 96
Pomace, 91
Portland cement, 200
Potato
  by-products
    flakes, 84
    proteins, 85
    pulp, 87
    starch, 80
    starch flour, 83
  dried, 82
  raw, 81
Poultry
  by-products
    hatchery waste, 164
    litter, 162
    offal meal, 163
  fat, 63
Protein concentrates
  alfalfa (PX1), 113
  fish, 144
  fodder protein, 140
  krill, 145
  milk (whey proteins), 186
  potato, 85
  soya-bean isolate, 123
Pulp
  dried, citrus fruits, 89
  grape, 93
  potato, 86
  sugar beet, 70
  tomato, 97
Pruteen ICI, 143

Q
Queensland arrowroot (or achira), 71

R
Rape
  full fat, 115
  hulls, 102

Rapeseed meal,
  dehulled, 125
  expeller meal, 126
  solvent-extracted meal, 127
Rice
  by-products
    bran, 40
    broken, 38
    shorts, 39
  paddy, 14
Rootlets (of barley), 37
Roots
  beet
    fodder-, 69
    fodder-sugar, 68
    sugar-, 67
    sugar-pulp, 70
  carrot, 72
  cassava
    pelleted, 76
    roots, 77
  common chicory, 74
  Jerusalem artichoke (or topinambur, fresh), 88
  rutabaga (or Swedish turnip), 87
  Swedish turnip (or rutabaga), 87
  topinambur (or Jerusalem artichoke, fresh), 88
  turnip, 78
Rutabaga (or Swedish turnip), 87
Rye (winter-), 16

S
St John's bread (or carob)
  fruit, 50
  germ, 51
Salt
  sea salt, 223
  sodiun chloride, 222
Scenedesmus (alga), 137
Sea salt, 223
Sea shells, 195
  oyster, 194
Shorts
  hard wheat, 19
  soft wheat
    red, 24
    white, 23
Single-cell proteins
  alga
    chlorella, 136
    scenedesmus, 137
    spirulina, 135
  Pruteen ICI, 143
  yeast
    brewer's-, 138
    distiller's-, 139
    fodder protein, 142
    lactic, 141
    torula, 140

Sodium
  caseinate, 179
  sources of
    bicarbonate, 224
    chloride, 222
    sea salt, 223
Soft wheat, 2
  by-products
    bran
      coarse, 26
      fine, 25
    germ, 22
    middlings, 21
    red shorts, 24
    white shorts, 23
Solubles
  maize DDGS, 27
  fish, 152
Sorghum
  high tannin, 7
  low tannin, 6
Soya-bean
  full fat, 122
  hulls, 114
  meal 44, 131
  meal 48, 132
  meal 50, 133
  proteins, 123
Spirulina (alga), 135
Starch
  maize, 46
  potato, 80
Starch flour (potato), 83
Sugar, 61
Sugar beet, 67
  fodder-, 68
  pulp of-, 70
Sugar industry (by-products)
  carbonate, 193
  concentrated vinasse, 99
Sugar cane (molasses), 59
Sunflower seed meal 34, 134
Swedish turnip, (or rutabaga), 87
Sweet potato, 79

T
Tallow, 66
Tomato
  kernels (or pips), 96
  pulp, 97
  skins, 95
Topinambur (or Jerusalem artichoke, fresh), 88
Tripolyphosphate (sodium-), 211
Triticale (French), 17
Tubers
  archira (or Queensland arrowroots), 71
  potato
    dried, 82
    flakes, 84

Tubers
  potato (*cont.*)
    proteins, 85
    pulp, 86
    raw, 81
    starch, 80
    starch flour, 83
  sweet potato, 79
  yam, 75
Turnip
  root, 78

V
Vegetable oils, 64
Vinasse
  concentrated
    sugar industry, 99
    yeast industry, 100
  standard, 98

W
Wheat
  buckwheat, 15
  hard-, 11
    bran, 20
    middlings, 18
    shorts, 19
  soft-, 2
    bran
      coarse, 26
      fine, 25
    germ, 22
    middlings, 21
    shorts
      red, 24
      white, 23
Whey by-products
  cow
    dried
      partially delactosed, 185
      mixed permeate, 184
      separated, mixed, 181
      separated, sweet, 180
    liquid
      casein-rennet, 172
      lactic casein, 173
      separated, acid, 169
      separated, mixed, 170
      separated, sweet, 168
  ewe, dried, sweet, 182
  goat, acid, 171

Y
Yam, 75
Yeast
  brewer's, 138
  distiller's, 139
  lactic, 141
  torula, 140
Yeast industry (concentrated vinasse of), 100